W9-APY-804

Date: 6/7/18

609 IKE
Ikenson, Ben,
Ingenious patents : bubble
wrap, barbed wire, bionic

PALM BEACH COUNTY
LIBRARY SYSTEM
3650 Summit Boulevard
West Palm Beach, FL 33406-4198

INGENIOUS PATENTS

BEN IKENSON AND JAY BENNETT

INGENIOUS PATENTS

Bubble Wrap, Barbed Wire, Bionic Eyes, and Other Pioneering Inventions

BLACK DOG
& LEVENTHAL
PUBLISHERS
NEW YORK

ACKNOWLEDGMENTS

First I need to thank my mother and father, Megan and Joe Bennett, my first and most reliable editors who instilled a sense of wonder in me when contemplating the workings of the world.

My editors for this project, Dinah Dunn and Hannah Smith, were essential to ironing out the details of the patents that we added to the book. And I must thank two editors at *Popular Mechanics*, Andrew Moseman and Eric Limer, who first introduced me to many of these technologies and continue to bring new marvels to my attention almost daily.

Finally, I would like to thank the U.S. Patent and Trade Office for fostering innovation, and sharing the patents of our greatest inventions with the world.

Contents

Chapter 4: The Strange and Sensational Patent   141

Chapter 5: The Patent That Eases   173

Introduction

"I admired the works of artists, but to my mind, they were only shadows and semblances. The inventor, I thought, gives the world creations which are palpable, which live and work." —Nikola Tesla

In centuries past, the patent was a certification of achievement—an inventor's final validation, in *writing*, that the creation of a device belonged to him or her. It represented hours of labor, a seemingly endless series of trial and error, and dozens, if not hundreds, of failures before the final invention came to fruition.

Today, patents fulfill a slightly different role. Yes, they can still be the final seal of approval for the rightful inventors of a new product, but the corporate use of the patent and the evolution of technology have changed things. Now, companies patent hundreds of devices, products, inventions, and ideas that they have not built and may never build, in an attempt to predict the future. The absurdity of patents ranges from a neck tattoo with a built-in microphone and lie detector, from Google, to a wheel that pops off your car so you can ride it like a scooter, from Ford. These exist so the companies can turn a profit just in case the inventions ever actually come to be.

Rarely do today's inventors tinker away in dark basements until the fruits of their labor finally, one day, after innumerable scalded hands and broken components, become something truly new to human civilization. But the Nikola Teslas of the world are not gone. They simply partake in a new process—a process of collaboration and evolution that defines technological advancement in the twenty-first century.

As a broad survey, *Ingenious Patents* celebrates all branches of the patent family tree, whether they stretch back centuries or are newly sprouted. The patents for Bubble Wrap and barbed wire can be just as intriguing as the patent for a bionic eye. A patent doesn't just outline an invention and describe how it works; patents are also a record of our values, our idiosyncrasies, and the spirit of invention that is fundamental to human nature.

Who invented the iPhone? Steve Jobs was the bold leader, but fourteen people are listed as inventors on the incredibly short, incredibly vague patent for an "electronic device." And really, the iPhone was not a *new* piece of technology but several pieces of existing technology—such as wireless connectivity, media storage, touchscreen user interfaces, an open platform for third-party apps, GPS—that culminated into what is arguably the most impactful product of the twenty-first century so far, changing not just how we live, but how we think.

What about the self-driving car? Big auto companies across the globe hold hundreds of patents related to autonomous vehicles, as do big tech companies. (Tesla, Inc., interestingly enough, holds none.) Yet it was a small Italian research firm, VisLab, that broke through the barrier, and in 2013 put the first car on public streets that was able to drive by itself. The future self-driving car—the one that will pick up your granddaughter from school—will be the result of endless tinkering, endless innovation, and vast stores of data acquired over decades and decades.

This illustrated collection of patents offers insight into some of the defining principles each invention represented, the inventor's original intention (sometimes wildly different from its ultimate use), and the peculiar visionary genius these patents were issued to protect. Many of the items were patented in an inventor's home country in addition to the United States. For the sake of consistency, this book relies mostly on the patent language and illustrations obtained from the United States Patent and Trademark Office.

Since Thomas Jefferson handed out the first patent in 1790, the U.S. Patent Office has granted more than nine and a half million patents in the effort to foster scientific advancement and economic prosperity. The inventions in these pages represent only a tiny fraction of these. In this revised edition of *Patents*, we have added fifteen new topics—3D Printer, Third-Generation Wireless Mobile Telecommunications (3G), Retinal Implant/Bionic Eye, Bluetooth, CRISPR Gene Editing, Helicopter Drone, Motorized Exoskeleton, Global Positioning System (GPS), Graphene, iPhone, Magnetic Levitation (Maglev), Self-Driving Car, Solar Panel, Virtual Reality, and Brain Implant—in an attempt to reveal the remarkable process of invention and innovation in the twenty-first century.

You probably didn't know, for example, that the quadcopter drone was invented in 1962. This new type of aircraft lay dormant for half a century until the technology around it—microprocessors and tiny cameras—revealed its full potential, and like the locusts of Africa's grasslands, consumer drones rose from the dust and took flight by the thousands.

Or take GPS, for example. Many inventions are the products of mistakes, accidental discoveries made while trying to build something completely unrelated, but GPS was more incidental than accidental. It was an afterthought—the leftover infrastructure of a national satellite program. Hardly anyone at the time could have predicted that so many people would carry powerful computers in their pockets everywhere they went, which was the hidden piece to the puzzle that created GPS as we know it today.

The inventions of the modern age are less likely to be wholly new devices that change the world, like lightbulbs, internal combustion engines, or airplanes. They are almost always single rungs up the ladder of civilization's advancement, completely reliant on each rung that came before. I hope this new edition, *Ingenious Patents*, will help reveal how high the ladder can extend.

—Jay Bennett

Patents: A History

"Congress shall have the power to … promote the progress of science and useful arts by securing for limited times to authors and inventors the exclusive right to their respective writings and discoveries." —Article 1, Section 8, U.S. Constitution

The Beginning

Although many of the United States' original thirteen colonies upheld some form of patent law, the original concept of the patent, as we now know it, was not a uniquely American idea. In 1449, King Henry VI of England awarded a patent to a "John" of "Utynam" for his distinctive way of manufacturing stained glass. It soon became clear that offering inventors some protection of their ideas benefited not only the individuals who came up with new things, but helped contribute to the economic growth of whole countries. So, when the settlers from England began arriving in America, they brought the concept of patent law with them.

So strong was the belief of the founding fathers in the usefulness of patent, trademark, and copyright law that tenets of the principle were written into the Constitution: Article 1, Section 8, Clause 8 of the U.S. Constitution, the "Intellectual Property Clause," is the basis upon which our patent laws are built. In fact, the contemporary patent office is founded on three acts passed during our country's early years: the Patent Act of 1790, the Patent Act of 1793, and the Patent Act of 1836.

Thomas Jefferson was among the group that led the charge to establish the first patent laws in the United States in 1790. The Patent Act of 1790 stipulated that all applications must be accompanied by a model of the invention, since Jefferson felt that patents should only be issued for tangible, physical things, not ideas. The first patent act also explicitly prohibited foreign patents from winning protection on U.S. soil. Samuel Hopkins of Pittsford, Vermont, was the recipient of the very first U.S. patent awarded by the new office, for his improvement in the making of "Pot Ash."

A few years later, a second act was passed, partly in response to the large number of complaints from inventors about the inefficiencies of the first act. In 1793, a compromise of two competing bills drafted by Thomas Jefferson and Alexander Hamilton was drawn up, resulting in the Patent Act of 1793. This act established a formal review board consisting of three positions, the secretary of state, attorney general, and secretary of war. However, there still was little organization within the patent office.

Reinventing the Patent Office

A rise in patent applications before the Civil War and during Reconstruction firmly established the need for a more efficient patent office. The Patent Act of 1836 established a patent office under the Department of State, and Henry Ellsworth, who helped draft the act, was appointed the first commissioner of patents. In addition to streamlining the application procedure, the Patent Act of 1836 also called for copies of all new patents to be distributed to libraries throughout the country, providing the general public with knowledge about the latest inventions. It is this act that the current patent office is based on.

In 1975, the patent office changed its name to the Patent and Trademark Office. Today it is now one of fourteen bureaus of the Department of Commerce and is located in Arlington, Virginia, where more than 5,000 people are employed by the office for the express purpose of examining patent and trademark applications.

Model Trouble

On December 15, 1836, a fire in the patent office destroyed all records and most of the patent models. Congress appropriated $100,000 for the restoration of 3,000 of the most important models. Forty years later, a second fire destroyed another 76,000 models. In 1880, the model requirement was deemed impractical, and in 1893, the remaining models were placed in storage.

Eventually, the models were sold at auction or became lost to obscurity. Some 2,500 of them ended up at the Smithsonian Institution, while thousands more landed in the estate of philanthropist Sir Henry Wellcome, who, it is believed, intended to establish a museum for them. Following his death, many of these were auctioned off. In 1979, Cliff Petersen, a designer and inventor within the aerospace industry, obtained some eight hundred crates—some of which had not been opened since they were packed up in 1926—for $500,000. Petersen donated 30,000 models and $1 million to the United States Patent Model Foundation, keeping approximately 5,000 models in his personal collection—yet many historic models were still landing in antique shops and flea markets.

A Home for Patent Models

In 1998, Alan Rothschild created the Rothschild Petersen Patent Model Museum in Cazenovia, New York. A significant segment of his collection was purchased from Cliff Petersen, and Rothschild also purchased all eighty-two models comprising the Patent Model Museum in Fort Smith, Arkansas. Thanks to the generosity of the Rothschild Petersen Patent Model Museum, several photographs of these models are represented in this book.

One

The Radical and
Revolutionary Patent

Army flight trials of the Wright airplane in Fort Meyer, Virginia, in July 1909

Airplane

Patent Name: Flying-Machine
Patent Number: 821,393
Patent Date: May 22, 1906
Inventor: Orville Wright and Wilbur Wright, of Dayton, Ohio

"Man must rise above the Earth—to the top of the atmosphere and beyond—for only thus will he fully understand the world in which he lives." —Socrates

What It Does   Enables human beings to travel by means of flight.

Background   On December 17, 1903, a telegram came from Kitty Hawk, North Carolina, that ushered in an important new chapter in history. Orville Wright informed his father of the success that he and his brother Wilbur had enjoyed that very morning when they tested their invention: the flying machine. On the last test of the morning, Wilbur had stayed airborne for fifty-nine seconds and traveled 852 feet.

The year 2003 marked a milestone for the airplane: the centennial of flight. On December 17, exactly a century after the Wright brothers succeeded, a reenactment was attempted—with unfortunate results. According to an Associated Press report, "a cheer rose from the crowd of 35,000 when the muslin-winged flyer roared to life and began moving down a wooden launch track. But that cheer suddenly turned to a groan when the rickety craft stopped dead in a muddy puddle at the end of the track."

To fly has been a dream of mankind since its beginnings; what took place that morning was, to a great extent, the culmination of exacting human will over countless eons. The Wright brothers had selected the Outer Banks of North Carolina for its windy atmosphere to help takeoff, and for the sands to soften their landing.

The Wright brothers' flying machine resembled a sophisticated version of a modern-day hang glider. Fastened into position, the operator propelled the apparatus forward along a rail; the biplane lifted off the ground propelled by both motor and wind. Its lightweight material and aerodynamic design were central to its success.

How It Works   The biplane design features two "aeroplanes"—parallel, superposed wings (1) and (2) in patent illustration figure 1—angled to such a degree as to be both

With a cruising speed of 1,350 mph, the Concorde was the world's fastest nonmilitary aircraft. It flew at supersonic speed on the edges of space, some 60,000 feet above ground. A trip from New York to London took just over three hours, less than half the flying time for a Boeing 747. Business trends, and the events of September 11, 2001, reduced market travel demand for this incredible aircraft; the Concorde dipped its nose for its farewell landing in late 2003.

propelled at a reasonable pace and lifted by wind. The aeroplanes are made of lightweight, sturdy, and flexible cloth-covered frames.

1. A series of transverse spars, longitudinal ribs, and diagonal threads forms a truss system to secure the planes together and provide a central component of the design.
2. V-shaped arms (23) help support a rear vertical rudder (22) attached by pivots (25), upon one of which is mounted a sheave or pulley (26).
3. Through the pulley, a tiller rope (27) passes, extending out laterally.
4. The tiller rope is secured to another rope (19) that passes along the top surface of the lower aeroplane, where shifting of the cradle (18) serves to turn the rudder.
5. At the front of the aeroplane, two struts (29) hold in place a horizontal rudder (31), the rear edge of which lies immediately in front of the operator and can be manipulated up or down.

In the Inventors' Words

"Our invention relates to that class of flying-machines in which the weight is sustained by the reactions resulting when one or more aeroplanes are moved through the air edgewise at a small angle of incidence, either by the application of mechanical power or by the utilization of the force of gravity.

"The objects of our invention are to provide means for maintaining or restoring the equilibrium or lateral balance of the apparatus, to provide means for guiding the machine both vertically and horizontally, and to provide a structure combining lightness, strength, convenience of construction, and certain other advantages.

"The apparatus is supported in the air by reason of the contact between the air and the under surface of one or more aeroplanes, the contact-surface being presented at a small angle of incidence to the air."

FIG. 1.

Model of an
artificial heart

Artificial Heart

Patent Name: Soft Shell Mushroom Shaped Heart
Patent Number: 3,641,591
Patent Date: February 15, 1972
Inventor: Willem J. Kolff, of Salt Lake City, Utah

What It Does

Assists or replaces the functions of a human heart.

Background

In 1982, an artificial heart was successfully transplanted into a patient who survived for 112 days following the procedure. Peter Houghton, who received the artificial heart in 2000, lived for seven years with the device, the longest to date of any patient.

Willem J. Kolff also invented an artificial kidney dialysis machine; he was inducted into the National Inventors Hall of Fame in 1985.

That the heart can be replaced successfully by an artificial substitute, even if only temporarily, is no less than a miraculous testament to human ingenuity. The blood vessels in an average human adult, if laid out in a single line, would wrap around the globe a few times. The heart is responsible for the distribution of blood through this sophisticated complex of veins and arteries. This is a big job that keeps getting bigger as we grow, and often more difficult if we risk heart failure from a variety of conditions, including chronic high blood pressure, coronary artery disease, diseased heart valves, or cardiomyopathy.

As doctors learn more about the heart's intricacies, the artificial heart continues to evolve. But the original brain behind the heart belonged to the Dutch-born American doctor Willem J. Kolff, a surgery professor at the University of Utah. Kolff received a patent for the first artificial heart on February 15, 1972. It happened to be the day after St. Valentine's Day, which happens to be Dr. Kolff's birthday: He was born on February 14, 1911.

How It Works

The artificial heart is a sophisticated device that works in two general cycles: filling and releasing.

In Figure 1:
1. A blood pumping chamber (10) is enclosed within a flexible wall (11).
2. A nonelastic, flexible net (12) restrains the wall from overdistending.
3. Blood enters the pumping chamber through an opening between a valve seat (15) and a valve head (14), atop the mushroom-shaped pumping member (13).
4. Another valve (16) prevents reverse flow of blood during the filling phase.

In Figure 2:

1. Air is forced into the mushroom-shaped pumping member.
2. While the pumping member expands, valve head (14) and valve seat (15) seal the chamber at the top, thereby preventing the backflow of blood.
3. Blood is then pumped out through the release valve (16).

In the Inventor's Words "It is desirable for the output volume of the artificial heart to be responsive to and substantially directly proportional to the blood input pressure…. The relationship between the output volume and inlet pressure is known as Starling's Law and is one of the important considerations in the design of an artificial heart, since this relationship will prevent the collapse of inlet blood vessels from an over suction by the pumping chamber or blood pooling from insufficient output volume."

Bluetooth

Patent Name: Contemporaneous Connectivity to Multiple Piconets
Patent Number: EP 1016242 A1
Patent Date: February 15, 2000
Inventor: Jacobus Haartsen, of Lund, Sweden

What It Does This wireless system allows electronic devices to exchange data over short distances using short-wavelength ultra-high-frequency radio waves.

Background Bluetooth, originally called short-link radio technology, was invented by Dutch electrical engineer Jacobus Jaap Haartsen in 1994 while he was working for Ericsson's Mobile Terminal Division. The technology allows electronic devices, such as smartphones, speakers, or a computer, to connect and share data without a cable by using low-power radio frequencies.

Although Ericsson has multiple U.S. patent applications related to Bluetooth technology, ongoing lawsuits and claims by patent trolls have muddied and delayed the application process for many of them. The term "patent troll" is used to describe a person or company that attempts to misuse patents to claim legal rights to an invention. The troll usually attempts to claim that older patents give them rights to a piece of new technology, even though the patents fall short of actually outlining the systems or mechanisms of that technology. Patent trolls are often able to delay the issue of legitimate patents for a technology and even achieve monetary gains from the ensuing lawsuits, as is the case with Bluetooth in the United States. Jaap Haartsen and Ericsson do, however, hold multiple European patents for key Bluetooth technologies, such as the one listed here.

Today, billions of devices worldwide use Bluetooth. Each electronic gadget that includes Bluetooth contains a small computer chip that serves as the Bluetooth radio and runs the necessary software to connect to another Bluetooth radio. When the chip is in close proximity to another Bluetooth device, the two can connect, or "pair," using radio frequencies between 2.4 and 2.485 GHz. Paired Bluetooth devices communicate with each other using short-range networks known as "piconets," which is the word used in Haartsen's patent. One device acts as the master and issues commands to up to seven connected devices.

Bluetooth technology has primarily been used for high-quality audio streaming, such as from a smartphone to a pair of Bluetooth headphones or a Bluetooth speaker. Bluetooth is used for much more than audio, however, as consumer products like smart basketballs and smart clothing also make use of the short-link radio technology. The ability to connect many electronic devices through low-energy radio waves has enabled the so-called Internet of Things (IoT), allowing multiple devices to work together, such as the thermostat, lights, locks, and even refrigerator in a smart home. The potential applications for Bluetooth are virtually endless as developers continue to find new uses and invent new devices to connect to a growing wireless network.

How It Works

1. A small computer chip in an electronic device serves as a short-wavelength ultra-high-frequency radio.
2. The computer chip runs software that allows it to broadcast its location to other Bluetooth devices, using radio waves with frequencies between 2.4 and 2.485 GHz.
3. When two Bluetooth devices recognize each other, they can connect or "pair," and wirelessly share data.
4. One Bluetooth device, often a smartphone or computer, serves as the master device, sending control commands to the other devices, such as headphones or speakers.
5. The wireless connection allows a multitude of applications, including streaming audio, connecting wireless keyboards to computers, tracking data from a fitness monitor, and even turning on your lights or locking your front door with your phone.

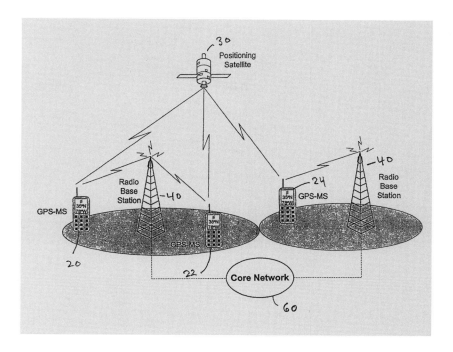

"I was asked to come up with a system that was wireless, digital and should be able to support both voice and data. I got a kind of blank sheet, and just made it. I looked at existing systems. I knew about the many mobile systems, but those were too heavy and too complicated for these really short-range technologies. I looked at cordless systems like DECT—digitally enhanced cordless telephony—but that was very much based on an infrastructure with base stations and cordless devices around it. And with Bluetooth, what I was asked to do was really a peer network, a point-to-point ad-hoc network between devices that were just in range."

Brain Implant

Patent Name: Three-Dimensional Electrode Device
Patent Number: US 5215088 A
Patent Date: June 1, 1993
Inventors: Richard A. Normann, Patrick K. Campbell, and Kelly E. Jones, of Salt Lake City, Utah

What It Does

This electronic device implanted in the brain communicates with neurons by sending and/or receiving electric signals. Neural implants can be used to monitor a subject's brain activity, restore senses that have been lost due to brain damage, such as vision and hearing, or even create a brain-computer interface.

A paralyzed woman drinks coffee using a brain implant to control a robotic arm.

Background Medical scientists in the 1870s realized that electrical stimulation of the brain could cause movements in both animals and humans. In the twentieth century, the first research was done to show that you could also alter a person's mood or behavior with electrical stimulation of the brain. In the early 1970s, Yale University physiologist José Delgado invented the "*stimoceiver*," a device implanted in the brain that stimulated the neurons with electric signals called nerve impulses. Delgado wrote in 1973 that stimulating the brain with radio signals "produced a variety of effects, including pleasant sensations, elation, deep, thoughtful concentration, odd feelings, super relaxation, colored visions, and other responses."

Another breakthrough occurred in neural implant technology in the late 1980s, when Richard A. Normann, Patrick K. Campbell, and Kelly E. Jones of the University of Utah invented the so-called "Utah Array"—an array of needle-shaped electrodes that could be implanted in the brain to stimulate specific neurons, or groups of neurons, with greater accuracy. A later patent application for a "brain implant device" that used a Utah Array is described, somewhat disturbingly, as "an implantable, integrated apparatus that contacts the brain with a plurality of metal needles to detect electrical signals or to transmit signals to the brain."

In recent years, brain implants have gotten much more sophisticated and nonintrusive, thanks to both wireless technology and ever-improving electrode arrays. Studies are still conducted with Utah Arrays, which are generally smaller than a dime, but other implants are beginning to replace them. For example, a "Stentrode" looks similar to a single sewing needle and can be placed within blood vessels on the side of the brain, and "neural lace" technology uses a mesh-like construction to cover larger parts of the brain in more electrodes for enhanced functionality. (Neural lace technology is thought to have the greatest potential to allow sophisticated control of a computer with the brain.)

Brain implants today can enable people to move and control robotic prosthetics or type out messages on a computer with their thoughts, and they can be used in conjunction with cameras and eye implants to restore a semblance of sight to the blind (see Bionic Eye). In March 2017, a fifty-six-year-old man paralyzed from the shoulders down in a cycling accident had doctors implant neural sensors in his brain and electrodes in his arm. The neural sensors detect his thoughts, or brain signals, and relay them to a computer, which sends signals to the electrodes in his arm via external cables, allowing him to move the limb.

"It was amazing," Bill Kochevar told the Associated Press. "I couldn't believe I could do it just by thinking about it."

In the future, brain implants could allow people to control even more advanced systems with their thoughts, such as programming a computer or driving a car. The medical benefits will improve as well, allowing doctors to bypass damaged neural connections to restore movement to paralyzed limbs.

The next big leap in neural implant technology is a device that could not only send signals to control computer systems but also receive signals from the computer to enhance the ability of the brain. Elon Musk, founder of Tesla, Inc. and SpaceX, recently announced a new company called Neuralink that will research and develop neural lace technology. These devices will first find applications in the medical industry, but Musk has expressed the opinion that using neural lace technology to enhance cognitive abilities will be essential for humans to contribute to a future economy run with artificial intelligence, saying at the 2017 World Government Summit in Dubai that we will need a "merger of biological intelligence and machine intelligence."

How It Works

1. A small electronic device, often including an array of electrodes, is implanted inside a person's head, either outside or within the skull, but most commonly the device is placed directly on the surface of the brain, attached to the cerebral cortex.
2. The neural implant can detect, block, record, or stimulate signals in groups of neurons in the brain's neural network.
3. Signals detected by the implanted device are then transmitted to a computer, either wirelessly or through a wired connection, and the computer system reads and interprets the signals.
4. The detected neural signals can then be recorded and mapped to develop models of the brain and to conduct research exploring how the brain works.
5. The signals can also be translated by the computer into a series of control commands that operate machinery, such as moving a robotic arm or prosthetic limb.
6. If the implant device includes an array of electrodes, then electric signals can be used to stimulate specific neurons in the brain and restore lost brain function or elicit emotional responses.
7. Researchers are working to develop brain implants that could bypass damaged nerve connections to restore functions such as movement or sight.
8. Some futurists believe that eventually, neural implants connected to computer systems could send signals to stimulate the brain and augment its functions—by increasing efficiency or improving memory, for example.

"This electrode array is specifically adapted to be used as a neuron interface device and to be implanted in the cortex of the brain. More specifically, the illustrated electrode array is adapted to be used as a vision prosthesis for a blind person. A visual image is reduced by some means, for example, a video camera, to an electrical signal that is then provided to the electrode array."

A Wolcott camera from 1840

Camera

Patent Name: Method of Taking Likenesses by Means of a Concave Reflector and Plates so Prepared as That Luminous or Other Rays Will Act Thereon
Patent Number: 1,582
Patent Date: May 8, 1840
Inventor: Alexander S. Wolcott of New York, New York

What It Does Captures live, still images on film, which can then be printed on paper and preserved as "photographs."

Background In 1840, Alexander S. Wolcott became the first person in the United States to receive a patent for a photographic invention. In Europe, Frenchman Louis-Jacques-Mandé Daguerre had already developed a process using light to capture images on plates. But these eponymous "daguerreotypes" required extremely long exposure times, so were best suited to inanimate objects. In 1839, Daguerre published a report on his process for producing daguerreotypes, and such creative and entrepreneurial Americans as Samuel F. B. Morse, who had been a portrait painter, and Alexander Wolcott seized upon the concept and set out to improve it. Wolcott, a dentist, came up with the idea of adding a concave

Daguerreotype of the inventor of the telescope, Henry Fitz, who partnered with Wolcott on the development of the camera

mirror inside the camera to reflect intense light onto a plate or film, significantly speeding up exposure time. Wolcott and a partner opened the world's first photo portrait studio in New York just two months before this patent was issued.

How It Works

1. A small open steel frame (C) stands on a support (D) within a box.
2. The support is fixed to a piece of wood (E) or other material that can slide along the bottom of the box.
3. A door at the top of the camera (A) serves as an eyepiece.
4. A concave reflector placed inside the box should reflect the image as it is desired to appear through the small steel frame.
5. The material on which the likeness will imprint is placed in a small steel frame.
6. By action of the light-reflecting mirror against the material, the image of the person whose likeness is being taken becomes imprinted on the material.

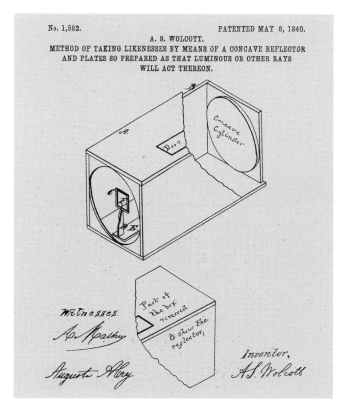

In the Inventor's Words

"When the camera—that is, the box A with the reflector, &c.—is to be used, the person whose likeness is to be taken should be placed in a chair, to which some suitable support for the head is attached to enable him to remain perfectly still. The camera should then be placed with the open end immediately opposite to the person. A trial plate is then to be placed or put against the frame C and the focus adjusted by sliding the piece E. The trial plate is then to be removed and the plate, paper, or other material to be used (prepared in any of the well-known methods for being acted on by luminous or other ways) put into its place and allowed to remain as long as required to form the image."

CRISPR Gene Editing

Patent Name: CRISPR-Cas Systems and Methods for Altering
Expression of Gene Products
Patent Number: US 8697359 B1
Patent Date: April 15, 2014
Inventor: Feng Zhang, of Cambridge, Massachusetts

What It Does

This genetic construct of DNA, RNA, and the Cas9 enzyme can be injected
into an organism to delete or replace parts of DNA molecules, changing the
organism's genetic makeup and physical characteristics.

Background

Feng Zhang, of the Broad
Institute, was the first to
successfully alter eukaryotes
using CRISPR technology. The
Broad Institute and University
of California, Berkeley are in a
legal dispute over who holds
the first patent.

CRISPR-Cas9 is a biomedical tool that can be used to edit the genes of living
organisms, altering the organisms' physical properties and/or immune-system
functions. The tool is a genetic construct, which is an artificially assembled
molecule of nucleic acids—the building blocks of a DNA strand. The entire
tool consists of three main parts: an RNA strand that can locate a specific segment of an organism's DNA; the Cas9 enzyme, which is capable of deleting that
segment of DNA; and a DNA strand that can be inserted to replace the deleted
strand and change the genome of the organism.

The technology has a number of applications, many of which are being re-
searched today in scientific studies that could lead to major commercial services
based on the tool. CRISPR-Cas9 can be used to engineer agricultural crops
with higher yields and greater resistance to diseases. It has been successfully
used in trials in China to create goats with twice the muscle and hair growth as
the unmodified species—which could augment the nation's livestock industry.
Research is even being conducted to modify pig organs with CRISPR-Cas9 so
they could be used in human transplants.

The primary focuses of CRISPR-Cas9 research, however, are related to
disease prevention. Scientists have used the gene-editing tool to engineer mosquitos that cannot harbor the malaria parasite, for example, with the long-
term goal of altering entire populations of mosquitos to reduce the dangers
of contracting malaria in high-risk countries. The United States and China
are also studying the possibility of using CRISPR-Cas9 to alter a human's
immune system so that its T cells are more effective at killing cancer cells.
This research could lead to entirely new treatments for cancer and possibly
even a cure. Research is also being conducted to eventually remove genetic mutations from embryos before birth, resulting in an entirely new way to prevent
genetic diseases before a patient is even born.

There were other gene-editing tools available to biomedical scientists before
CRISPR-Cas9, but this tool has proven much easier to use, more accurate, and
significantly less expensive. That said, as the technology continues to mature,
there are still risks that the wrong segment of an organism's DNA could be
deleted, or that a replacement strand could fail to take hold in the organism's
genome, causing unintended genetic mutations.

The University of California, Berkeley, and the Broad Institute—a research nonprofit partnered with Harvard and MIT—have entered into a legal dispute over who holds the first patent that covers CRISPR technology. The patent referenced here, awarded to Feng Zhang of the Broad Institute, was the first patent issued that directly details a successful experiment to alter eukaryotes, or multicellular organisms, which are the type of cells in humans. While Berkeley's patent was filed first and is based on earlier research, it outlines a simpler process to use CRISPR-Cas9 on prokaryotes, or unicellular organisms, and, specifically, bacteria. The method detailed by the Broad Institute most directly applies to the developing industries and continuing research that will use CRISPR-Cas9.

How It Works

1. A DNA construct is injected into an organism, often an embryo, so that the genetic changes will be applied to the entire organism, but in humans one can also withdraw living cells, edit them with CRISPR, and then inject them back into the patient.
2. An RNA strand, the genetic material that reads and decodes DNA, identifies the segment of DNA in the organism that is to be changed.
3. The Cas9 enzyme deletes that segment of DNA.
4. As an optional final step, a new DNA strand can be inserted where the old one was deleted. This is not required if the gene-editing tool is applied only to remove a trait, such as a genetic mutation, from an organism.

In the Inventor's Words

"For a long time now, we have known [the] genetic mutations that underlie disease. And so one of the thoughts is—if we know what mutations cause disease, why not just go in and replace it and revert it back to the normal sequence?"

DNA Fingerprinting

Patent Name: Method of Characterizing Genomic DNA
Patent Number: 5,175,082
Patent Date: December 29, 1992
Inventor: Alec J. Jeffreys, of Leicester, United Kingdom

What It Does

A person's DNA pattern is as distinctive as a fingerprint. Being able to create a genetic "fingerprint" from a DNA sample provides a definitive means of identification that can be used for paternity and maternity testing, in criminal investigations, or in forensic medicine.

Background

In 1992, DNA fingerprinting confirmed that the remains of a body found in Brazil in 1979 belonged to Dr. Josef Mengele, the notorious "Angel of Death" of the Auschwitz death camp.

Sir Alec Jeffreys invented a genetic technique that has revolutionized criminal investigation and forensic science. DNA, or genetic, fingerprinting, involves a technique used to distinguish individuals within a given species based on their unique DNA.

In September 1984, Jeffreys, a scientist and professor at the University of Leicester, England, was studying the evolution of genes at the British Antarctic Survey Headquarters. To track and mark the positions of genes, it was useful to isolate small areas of DNA where high levels of variability were found. Jeffreys was using a lump of seal meat as his subject, comparing the seal gene with the closest counterpart gene in a human.

His comparisons led him to devise a method of detecting lots of variations in genetic material simultaneously, leading in turn to a surprising revelation. The DNA sequences in his genetic studies soon came to reveal signatures far more specific than those which distinguish species.

Jeffreys and his colleagues had hit upon the first DNA "fingerprint," and they knew right away that its implications were huge. A foolproof system of identification for establishing family relationships, investigating crime, conducting medical research, and even studying wildlife could have far-reaching applications. In 1994, the science that Jeffreys helped spawn made a very public appearance: Prosecutors obtained and presented DNA evidence that put O. J. Simpson at the scene of a double murder. Now that DNA evidence is routinely used in criminal investigations, a number of people serving time for capital crimes have been exonerated and freed because the DNA evidence pointed to someone else.

Conventional fingerprinting preceded the more exacting DNA tests that now aid in identification.

How It Works

Every cell in the human body contains a copy of a DNA blueprint made up of twenty-three pairs of chromosomes. Only a tiny portion of the human genome varies from one individual to another. Methods of DNA fingerprinting focus on these small chromosomal regions of variability. Here DNA fragments reveal paired chromosomal sequences—variable number tandem repeats, or "microsatellites"—that repeat anywhere from one to thirty times in a row. Two different numbers of repeats in a single sequence come from the paired paternal and maternal chromosomes. These numbers, probed at various variable number tandem repeat locations, create a statistically unique DNA profile.

In the Inventor's Words "The present invention relates generally to polynucleotides and DNA and RNA probes, their preparation and their use in genetic characterization. Such uses may include, for example, establishing human, animal or plant origin, and the polynucleotides and probes of the invention may thus find use for example in paternity disputes or forensic medicine or in the prevention, diagnosis, and treatment of genetic disorders or predispositions.

"In British patent application No. 8525252 (publication no. 2166445) there are described various DNA sequences which may be used as probes to hybridize individually at a number of polymorphic sites within the human and animal genomes enabling the production of a 'fingerprint' composed of marked bands of differing molecular weights. The fingerprint as a whole is characteristic of the individual concerned and the origin of the differing bands can be traced through the ancestry of the individual and can in certain cases be postulated as associated with certain genetic disorders."

Dynamite sticks primed for explosion

Dynamite

Patent Name: Improved Explosive Compound
Patent Number: 78,317
Patent Date: May 26, 1868
Inventor: Alfred Nobel, of Stockholm, Sweden

What It Does This relatively safe and effective explosive (contained in rods, stable to transport, and able to be detonated from a distance) can create controlled, focused explosions of various intensities, for the purpose of demolition.

Background

The chemical explosive nitroglycerine was discovered by Italian chemist Ascanio Sobrero in 1846.

The 1911 Nobel Prize in Chemistry was awarded to Marie Curie, the French chemist who discovered the elements radium and polonium. Her findings would soon contribute to the creation of an explosive force far more powerful than nitroglycerine.

Alfred Nobel was both a scientist and a worldly entrepreneur. The son of a wealthy inventor and engineer of bridges, he understood how a little ingenuity could be applied to turn a large profit and change the way work is done. To these ends, he came up with dynamite. One of the most important tools of the era, dynamite instantly made mining, construction, and highway and railroad building faster and cheaper; and it contributed significantly to the Nobel family fortune.

Before his invention was patented, Nobel worked with his father to develop nitroglycerine as a commercially viable explosive for blasting rocks. Experimenting with nitroglycerine, an oily fluid, proved a dangerous business that did not come without human cost. One early explosion killed several people, including the inventor's brother. But Nobel remained determined to succeed, and tinkered with organic additives in order to make the handling of nitroglycerine safer. By mixing silica with nitroglycerine, he found that the liquid could be made into a malleable paste. The paste could then be shaped into rods. The rods could be inserted into holes drilled where explosions would be safest and most strategic.

During his lifetime, Nobel held more than 350 patents in different countries in a variety of fields, including electrochemistry and physiology. He willed much of his fortune to establish a fund that awards achievements in physics, chemistry, physiology and medicine, literature, and peace. The annual Nobel Prize ceremony is held in Stockholm, Sweden, where Nobel was born.

How It Works

Nitroglycerine and absorbent silica are mixed together at a ratio of approximately 60/40 for a minimally effective explosion, up to about 78/22 for a more powerful blast.

1. Confined in a tight, strong enclosure, the compound is detonated by the application of heat above the temperature of 360° Fahrenheit.
2. Ignition is accomplished by the use of a fuse or detonator, which can be activated from afar.

In the Inventor's Words

"So great is the absorbent quality of this earth, that it will take up about three times its own weight of nitroglycerine and still retain its powder-form, thus leaving the nitroglycerine so compact and concentrated as to have very nearly its original explosive power; whereas, if another substance, having a less absorbent capacity, is used, a correspondingly less proportion of nitroglycerine will be absorbed, and the powder be correspondingly weak or wholly inexplosive."

Alfred Nobel

GPS

Patent Name: Navigation System Using Satellites and Passive Ranging Techniques
Patent Number: US 3789409 A
Patent Date: January 29, 1974
Inventor: Roger L. Easton, of Craftsbury, Vermont

What It Does
This global navigation satellite system, operated by the U.S. Air Force and paid for by American taxpayers, provides specific location and time information to any device with a GPS receiver.

Background
Today, GPS is an everyday technology on most mobile devices and computers. Applications that use it, like Google Maps, can pinpoint your location and provide directions to another location. The story of GPS, however, is the story of the U.S. satellite program, as GPS was the result of the natural evolution of technologies used to track satellites, and methods of using satellites to track objects on Earth.

Roger L. Easton was the mastermind behind the primary technologies used in GPS. He cowrote the Naval Research Laboratory (NRL) Project Vanguard proposal in 1955, which was ultimately selected by the Eisenhower administration to launch some of the United States' very first satellites. In 1957, Easton invented a system to track the Vanguard satellites, called Minitrack, and he expanded that system to track unknown satellites after the Soviet Union launched Sputnik 1 later that year.

Easton went on to design the Naval Space Surveillance System in 1959, the first radar network to locate and track any orbital object passing over the United States. In the late 1960s and early 1970s, Easton developed the technologies needed to locate and track objects on Earth from space, as opposed to his previous systems that tracked objects in orbit from the ground. Specific orbital trajectories were established, and high-precision clocks were launched in satellites, such as Timation 1 and 2, to design the Global Positioning System. In 1977, Navigation Technology Satellite 2 became the first satellite to transmit GPS signals.

Due to special and general relativity, it appears to observers on Earth that the clocks on board the orbiting GPS satellites run about 38 microseconds per day faster than clocks on the ground. This inconsistency could create distance errors in GPS data up to 10 kilometers, which would continue to get worse over time. Easton, along with Ivan A. Getting of the Aerospace Corporation and Bradford Parkinson of the Applied Physics Laboratory, is credited with designing the system that corrected for the time differences, to provide accurate global positioning.

Roger Easton

GPS became fully operational in 1995 with a network of twenty-four satellites. Easton was awarded the 2004 National Medal of Technology by President George W. Bush, and he was inducted into the National Inventors Hall of Fame in March 2010.

How It Works

1. The GPS network is made up of three main parts: satellites, radar stations on Earth, and GPS receivers.

2. The exact location of all satellites in a network of more than thirty is known at any given time based on their orbits.

3. Radar stations on the ground are used to make sure the satellites are in fact where predicted, and correct for any variations.

4. A receiver in an electronic device, such as a smartphone, receives radio signals from these satellites as they fly overhead, and can determine exactly how far away the device is from the satellite.

5. The GPS receiver must be locked on to the signal of at least three satellites to determine latitude and longitude, and to track movement. A fourth satellite can determine the altitude of the receiver, and if they are available, multiple satellites can be used to refine the data.

6. The geolocation data is transferred to a map display on the electronic device.

FIG. 1.

In the Inventor's Words "Until recently, celestial navigation techniques have provided mankind with the best available accuracy for terrestrial navigation. A new order of accuracy has become possible with the introduction of artificial earthsatellite navigation systems wherein the instantaneous location of the satellites is precisely known."

Graphene

Patent Name: Nano-Scaled Graphene Plates
Patent Number: US 7071258 B1
Patent Date: July 4, 2006
Inventors: Bor Z. Jang and Wen C. Huang, of Dayton, Ohio

What It Does This two-dimensional sheet of carbon molecules, assembled in a hexagonal lattice on the atomic scale, can be layered to form incredibly strong composite materials that conduct both heat and electricity.

Background Graphene is a single layer of carbon molecules, just one atom thick, that is composed of hexagonal shapes arranged in a honeycomb structure. It comes from chunks of graphite, such as you find in a pencil, though it is very difficult to isolate a single layer. The first images of a small number of graphite layers were taken with a transmission electron microscope and published in 1948.

Throughout the twentieth century, scientists worked to study thinner and thinner layers of this intriguing material.

It wasn't until 2004 that Andre Geim and Konstantin Novoselov at the University of Manchester extracted single-layer one-atom-thick crystals of graphene. They did so using a method called the "Scotch tape technique," which uses an adhesive tape to peel a layer of graphene from a chunk of graphite and deposit it on a substrate of silicon dioxide. Geim and Novoselov were awarded the Nobel Prize in Physics in 2010 "for groundbreaking experiments regarding the two-dimensional material graphene."

Eight years earlier, in October 2002, Bor Z. Jang and Wen C. Huang filed U.S. patent 7071258 B1. The patent, granted in July 2006, describes one of the first methods for extracting graphene, an exfoliating technique that produces thin layers. "Preferably, all length, width and thickness values are smaller than 100 nm," reads the patent. Geim and Novoselov's technique, however, produced layers just an atom thick, which is less than half a nanometer.

Graphene has many intriguing properties, giving the material a large number of potential applications. Layers of graphene are incredibly lightweight while maintaining high strength—about two hundred times stronger than layers of steel of the same thickness. Manufacturers and researchers are working to find an efficient way to develop graphene structures on a large scale.

Thanks to the material's heat-resistant qualities, electrical conductivity, and incredibly high melting point—about 4900 Kelvin—graphene is being considered as a replacement for silicon in computer chips. Graphene-based transistors can transport electric currents tens of times faster than silicon, meaning computers could run faster with less power. The material has also been used to manufacture high-efficiency solar panels and fast-charging batteries. Research related to applications for graphene is widespread, from medical diagnosis devices to optical lenses, to water filtration systems and airframe manufacturing. As fabrication methods improve, graphene will become more widespread in electronic devices and manufactured goods.

How It Works
1. Mechanical cleavage, also known as the "Scotch tape technique," uses an adhesive film to peel a single layer of graphene from a larger chunk of graphite.
2. To produce graphene in larger quantities, additional chemical processes can be used.

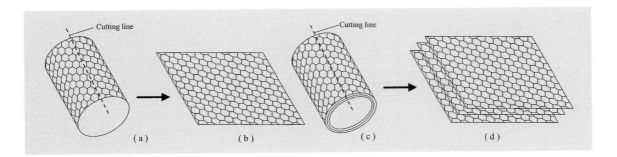

(a)   (b)   (c)   (d)

Cutting line   Cutting line

3. After extracting graphite flakes, scientists suspend the crystals in a solution.
4. The graphite flakes are oxidized with sulfuric or nitric acid, which inserts oxygen atoms in between the individual layers of graphene and forces them apart.
5. The solution with the suspended graphene layers is then filtered through a membrane with pores that are about 25 nanometers wide.
6. The filtering process is controlled, so that when a crystal of graphene covers up one of the pores, the liquid is directed to drain through other unclogged pores.
7. By covering all of the pores in the membrane, uniform single layers of graphene are deposited.
8. The graphene is then applied to a substance, such as glass or plastic, and the membrane is washed away.
9. The film is finally treated with the chemical hydrazine to convert the oxidized graphene into pure graphene.

In the Inventors' Words   "I would say there are three important things about graphene. It's two-dimensional, which is the best possible number for studying fundamental physics. The second thing is the quality of graphene, which stems from its extremely strong carbon-carbon bonds. And finally, the system is also metallic."

Igor Sikorsky in an early helicoptor

Helicopter

Patent Name: Direct-Lift Aircraft
Patent Number: 1,994,488
Patent Date: March 19, 1935
Inventor: Igor I. Sikorsky, of Nichols, Connecticut

What It Does   This compact, versatile, rotor-powered flying machine is capable of taking off and landing without the need for a runway, moving in multiple directions at variable speeds, flying close to the ground, and carrying passengers and cargo for relatively long distances.

Background

Although he did not invent it, Igor Sikorsky is considered the father of the helicopter. He was inducted into the Inventors Hall of Fame in 1987.

"The work of the individual still remains the spark that moves mankind ahead."
—Igor Sikorsky

Igor Ivan Sikorsky was one of those individuals who seemed to find his path through life at an early age. Born in Kiev, Russia, he was fascinated by flight as a child; in his early twenties, he was a leading figure in Russian aeronautics. In 1913, when he was just twenty-four, Sikorsky built the world's first four-engine airplane, a close ancestor to the bomber airplanes of World War I.

Pursuing his dreams, Sikorsky moved to the United States and established the Sikorsky Aero Engineering Corporation. From the mid-1920s to 1940, Sikorsky built a variety of large passenger-carrying amphibious aircraft. His "flying boats" garnered him success and fame, and, more importantly to him, enormous credibility in the realm of U.S. aeronautics.

All along, Sikorsky had been interested in creating a successful helicopter, and worked on the invention of a direct-lift single-rotor aircraft. Back in Russia when he was twenty, he had tried to build one, to no avail. But thirty years later, his determination bore fruit: On September 14, 1939, the VS-300 made its first flight. The first commercially successful helicopter has since become a prototype for single-rotor helicopters, the world's most versatile aircraft, and Sikorsky is viewed as one of history's most important contributors to the field of aeronautics.

How It Works

Unlike airplanes, helicopters can fly backward, hover in midair, and rotate, and require only a minimum of space for takeoff and landing.

1. A single rotary propeller at the top of the helicopter creates the necessary lift and can be tilted to maneuver the vehicle.
2. A smaller rotary propeller at the tail offers a steering mechanism.
3. While it seems relatively simple, operating a helicopter requires a great deal of coordination and practice, and the use of both feet and both hands.
4. One hand control is set to operate the lateral direction of the vehicle, which tilts the main propeller.
5. Another hand control is set to operate the vehicle's vertical direction by adjusting the speed of the main propeller.
6. Two foot pedals control the tail rotor, which acts as a rudder for the vehicle.

"The object of my invention is to solve successfully such direct lift aircraft problems as torque compensation, directional steering, and the application of power to a vertical lift producing propeller, either by a power plant or by means of air through which the vehicle navigates.

"A further object of the invention is the creation of a cheap, simple and easily operated mechanism for controlling the lateral and longitudinal stability of the aircraft and guiding it positively in its various directions of flight."

Internal Combustion Engine

Patent Name: Gas-Motor Engine
Patent Number: 365,701
Patent Date: June 28, 1887
Inventor: Nikolaus August Otto, of Deutz-on-the-Rhine, Germany

What It Does

Provides an apparatus for igniting gas-motor engines in which the charge is subject to compression before ignition. Otto's engine is the prototype of the combustion engine. The piston draws in and compresses a gas-air mixture inside a closed container. A spark ignites the mixture, and it explodes.

Background

Fellow countryman and close colleague Gottlieb Daimler is believed to have used Otto's engine in building the world's first motorcycle in 1885. Daimler had been technical director at the company Otto co-owned and later went on to modify his own version of the gas engine.

The world's love affair with the automobile began as soon as the first Model T Fords started rolling down the streets, and it's been gaining momentum ever since. Yet, one of the most important innovations in automobile history was invented by someone who was never actually involved in making cars. Nikolaus Otto was a traveling salesman before he became interested in designing engines. He started an engine manufacturing company with a partner in 1864. In 1876, he devised the first practical four-stroke internal combustion engine.

Otto's company sold over 30,000 of the new engines over the next decade. Now called Deutz AG, the company Otto cofounded is the oldest manufacturer of internal combustion engines. The "Otto-Cycle Engine," as it was named in honor of its inventor, remains a model for engines made today. Indeed, on the long road of automobile history, the internal combustion engine was a major mile-marker, if not a true starting point.

How It Works

A piston is employed to facilitate the following four strokes of an internal combustion engine:

1. An intake valve opens to draw a mixture of air and gas into the cylinder.
2. The mixture is compressed in the cylinder.
3. A small gas charge explodes in the cylinder.
4. A valve opens to release the exhaust.

In the Inventor's Words

"This invention relates to an improved construction of apparatus for igniting the charges of those gas-motor engines in which the charge is subject to compression before ignition.

"The apparatus consists of a slide that is pressed against a slide-face of the cylinder by means of a loose cover acted on by springs, and is worked by suitable gearing from the engine-shaft, such slide being provided with posts and passages arranged and operating as I will proceed to explain with reference to the accompanying drawing, in which—

"Figure 1 shows a horizontal section of the apparatus at the moment of igniting the cylinder-charge. Figure 2 shows an end elevation. Figure 3 shows a view of the inner face of the slide. Figure 3ª shows a cross-section; and figure 4 shows the same view as figure 1, with the slide at its extreme right-hand position."

CHRONOLOGY OF THE AUTOMOBILE AGE

Perhaps no other invention has changed the lives of so many as the automobile. The prevalence and ease of use of the gasoline engine have given people a freedom of movement that has forever changed the face of the global landscape. But no one person can claim the honor of having invented the automobile as we know it today: The gasoline-powered car is the result of numerous innovations throughout history, beginning as far back as the eighteenth century.

1769–70: The first non-horse-drawn road vehicles were built by Nicolas-Joseph Cugnot, a French military engineer. He built a passenger vehicle and a cargo tractor, both of which were powered by steam. However, steam-powered vehicles were soon deemed too dangerous, dirty, and noisy for regular street use, and engineers started looking for new ways to power vehicles.

1860: Jean-Joseph-Étienne Lenoir, a Belgian-born Frenchman, patented the first viable internal combustion engine. Similar to a steam engine, the single-cylinder engine was fueled by streetlight gas and ignited by a storage battery, and was used mainly to power machines. But his invention proved too unwieldy for everyday use. After selling about five hundred engines, Lenoir died in poverty.

Adolph Daimler, the son of Gottlieb Daimler, and his wife, Mercedes, in the Mercedes 35 PS in 1901

◀ **1885:** Gottlieb Daimler and Karl Benz, both engineers in Germany, separately worked on developing engines based on principles of internal combustion. Daimler later went on to found the Daimler Motor Company in 1890, and unveiled the first Mercedes in 1901.

1887: German engine manufacturer Nikolaus Otto patented his "Gas-Motor Engine," a revolutionary invention in automobile history. It was the first commercially successful four-stroke internal combustion engine; it remains a prototype for today's engines.

1891: Long before Toyota introduced the Prius, American inventor William Morris unveiled the first electric car, powered by batteries that were stowed under the car's seats. Though quiet and nonpolluting, it required frequent recharging and couldn't go very fast. Its initial popularity soon died.

1893–94: Brothers Charles E. and J. Frank Duryea introduced the first gasoline-powered automobile in the U.S., and later founded the Duryea Motor Wagon Company, the first U.S. company to produce gas-powered automobiles.

1895: Michelin, a French company, introduced rubber tires filled with compressed air, giving automobiles their first smooth ride.

1896: Henry Ford built the first successful gasoline-powered car, an early prototype of the Model T.

1901: Huge oil fields were discovered in eastern Texas, making gasoline cheaper and more accessible, and helping accelerate the popularization of the automobile.

The first Oldsmobile in 1897

◀ **1901:** Ransom E. Olds began mass-producing his namesake car, the Oldsmobile. It featured a curved dashboard, injecting the automobile with a sense of style.

1903: Henry Ford founded the Ford Motor Company in Michigan.

1904: Henry M. Leland's Cadillac Automobile Company began manufacturing cars with interchangeable parts, helping to streamline the production process.

Ernerst A. Franke drives his 1921 Model T to the White House to show Henry Ford, who was visiting President Roosevelt in 1938.

1908: Henry Ford unveiled the iconic Model T, which came to be known affectionately as the Tin Lizzie. The Model T outsold all other cars for the next twenty years.

1911: Charles F. Kettering invented the electric self-starter. This ingenious invention allowed drivers to start their cars electronically, rather than using the cumbersome hand cranks that were until then standard.

1913: Henry Ford's automobile factory installed a moving assembly line, increasing the speed at which cars could be manufactured, decreasing the cost of the final product, and making cars more affordable for the masses.

1914: U.S. automakers agreed to cross-licensing, the system of sharing patents without having to pay one another for the information. This collaborative spirit helped accelerate innovation in automobile manufacturing in the early part of the twentieth century—but inevitably, the system of cross-licensing ended in 1956.

1918: Malcolm Loughead invented the hydraulic four-wheel brake system, making it easier and safer for drivers to bring their cars to a halt.

1922: Low-pressure tires, known as balloon tires, were introduced. The decrease in pressure made riding in a car more comfortable.

1939: Technological improvements contributed to the ease of operating an automobile and driving long distances. The fully automatic transmission decreased physical exertion by eliminating the need to switch gears manually, while air-conditioning systems helped control the comfort of the car's interior.

1956: The Federal-Aid Highway Act of 1956 was introduced under President Dwight D. Eisenhower. It earmarked federal funds for the completion of a nationwide network of highways in the United States.

1968: In order to help control environmental pollution, the U.S. mandated that automobiles be equipped with devices to reduce exhaust fumes.

1975: The worldwide oil shortage led the U.S. Congress to pass a law requiring that cars be made more fuel-efficient.

1997 and beyond: Toyota introduced the Prius, a mass-produced hybrid car that combines the gas engine with an electric motor. Honda followed suit in 2000 with its hybrid car, the Insight. Auto manufacturers began exploring alternate sources of fuel energy, such as the fuel cell, which burns hydrogen as fuel, producing only water as a by-product. The car of the future will certainly be less polluting and more fuel-efficient than ever before.

iPhone

Patent Name: Electronic Device
Patent Number: US D672769 S1
Patent Date: December 18, 2012
Inventors: Bartley K. Andre, Daniel J. Coster, Daniele De Iuliis, Richard P. Howarth, Jonathan P. Ive, Steve Jobs, Duncan Robert Kerr, Shin Nishibori, Matthew Dean Rohrbach, Douglas B. Satzger, Calvin Q. Seid, Christopher J. Stringer, Eugene Antony Whang, and Rico Zorkendorfer, of Cupertino, California

What It Does

This touchscreen phone also functions as an Internet-browsing device and a media player—a handheld computer.

Background

Steve Jobs

One of the vaguest patents ever issued, simply for an "electronic device," is actually the patent for the original iPhone. Though the iPhone was introduced by Steve Jobs in 2007 at San Francisco's Macworld Expo, the patent was not issued until December 2012. The patent includes eight diagrams, but the only written description is: "The ornamental design of an electronic device, as shown and described." Also included is a long list of patents for technology used in the iPhone, a handful issued to Apple, and eight simple diagrams that show the general shape and structure of the twenty-first century's most revolutionary piece of consumer tech thus far.

The first iPhone was primarily made of aluminum, and while the materials and internal electronics have been upgraded, the general design of the device has remained largely unchanged in a decade. Each iPhone since the first generation has retained its finger-driven touchscreen, a home button on the bottom of the screen (though this will likely be abandoned in future iPhones), a camera on the back, a speaker, a microphone, and volume controls on the left side. The original iPhone was designed to roll the MP3 player, a phone, and Internet access into one device. As the phone has evolved and grown in capability, with an entire industry sprouting up to develop iPhone apps, the device is most accurately described as a handheld computer.

Many additional patents related to the iPhone and its primary functions have been issued. U.S. patent 7479949 B2, issued to Steve Jobs and company in January 2009, describes a touchscreen interface based on a heuristic process—meaning the interface is designed so users can easily teach themselves how to operate the device. Rational controls, such as swiping across the screen to navigate different pages and using two fingers to zoom in or out, are outlined in this patent. The idea of building a device that anyone could pick up and instinctively learn to use became the driving ideology behind the family of iPhone and iPad devices.

In July 2016, before the release of the iPhone 7 in September of that year, Apple sold its one-billionth iPhone. As of 2016, there were about 2.6 billion smartphone users out of the 7.4 billion people in the world. That number is expected to increase to over 6 billion by 2020, more than 75 percent of the projected world population for that year. Apple's iOS devices and devices that run Google's similar Android operating system comprise the vast majority of smartphones worldwide, and the ubiquitous use of smartphones has fundamentally changed how our species navigates, communicates, acquires information, and even thinks.

How It Works

1. The iPhone uses a variety of electronics powered by a lithium-ion battery and controlled via a touchscreen-based interface.
2. The heart of the device is a system on a chip, or SoC, that integrates all of the components of a computer into a single chip, including multiple processor cores, memory blocks, and connecting circuits.
3. Multiple electronic components are controlled via the CPU, or central processing unit, including a display screen, digital camera, GPS electronics, ports for wired connections, and wireless, Bluetooth, and cellular connectivity hardware.
4. Apps developed by third parties can be downloaded via Apple's App Store to increase the functions of the device with new programs.
5. The user interface stems from the touchscreen, where app icons can be selected and an electronic keyboard can be used to type.

In the Inventors' Words

"In 2001, we introduced the first iPod, and… it didn't just change the way we all listen to music, it changed the entire music industry. Well, today, we're introducing three revolutionary products of this class. The first one: is a widescreen iPod with touch controls. The second: is a revolutionary mobile phone. And the third is a breakthrough Internet communications device. So, three things:

a widescreen iPod with touch controls; a revolutionary mobile phone; and a breakthrough Internet communications device. An iPod, a phone, and an Internet communicator. An iPod, a phone...Are you getting it? These are not three separate devices, this is one device, and we are calling it iPhone. Today, today Apple is going to reinvent the phone, and here it is."

Thomas Edison
in 1880

Lightbulb

Patent Name: Electric-Lamp
Patent Number: 223,898
Patent Date: January 27, 1880
Inventor: Thomas A. Edison, of Menlo Park, New Jersey

What It Does

Provides light by means of a glowing filament inside a glass globe, powered by electricity.

Background

On September 4, 1882, the first commercial power station began providing light and electricity to customers in a one-square-mile area of Lower Manhattan.

Thomas Edison was a genius credited with a great number of inventions, but the electric lightbulb was not entirely his own creation. Many inventors throughout the world were trying to replace the gas lamp with some sort of lightbulb. In fact, Edison's British contemporary Joseph Swan beat him to the punch when Swan demonstrated the world's first electric lightbulb in 1878. A few men before this—the Canadian Henry Woodward and his partner Matthew Evans— patented a lightbulb, but they didn't have the capital necessary to commercialize their invention. Edison, who was well established by then, purchased the rights to their patent.

But Edison was not simply buying his way into another invention. He worked at breakneck speed to invent a variety of components that would ensure that his light would shine with the brightest commercial success.

Ultimately, Edison ran his new lamp for two days and forty minutes. The day that light burned out is the date usually given for the first commercially practical lamp—October 21, 1879. By 1880, Edison had perfected a 16-watt incandescent bulb that could burn for up to 1,500 hours.

How It Works

Incandescent lightbulbs—those which emit light as a result of heat—are less efficient than more recent innovations: neon, fluorescent lighting, and LEDs (light-emitting diodes). Conventional lightbulbs use significantly more energy heating the filament, wasting electricity in the process. A 60-watt incandescent lightbulb lasts about 750 hours, about ten times less than a compact fluorescent bulb that generates the same amount of light.

1. Glass is blown into a bulb-shaped mold.
2. A thin wire or filament is drawn out and wound into a double coil.
3. At both ends, the filament is attached to power leads embedded in a supporting glass structure.
4. This support is inserted into a bulb and the two glass parts fused together.
5. A tube in the support is used to replace the air inside of the bulb with inert gases.
6. The tube is cut, the opening is sealed, and the base of the bulb is attached.
7. Electric currents flow from one contact to the other, traveling through the wires and heating the tungsten filament, which in turn emits light.

"Be it known that I, THOMAS ALVA EDISON, of Menlo Park, In the State of New Jersey, United States of America, have invented an Improvement In Electric Lamps, and in the method of manufacturing the same (Case No.186) of which the following is a specification.

"The object of this invention is to produce electric lamps giving off light by Incandescence, which lamps shall have high resistance, so as to allow the practical subdivision of the electric light."

INVENTOR PROFILE

Thomas A. Edison (1847–1931)

Thomas Edison experimenting in his laboratory in the 1920s

No book about patents would be complete without a profile of Thomas Alva Edison. Edison was awarded 1,093 patents—the greatest number of patents the U.S. Patent Office has ever given to a single person. In addition, he received thousands more patents from dozens of other countries. His inventions helped advance the technology of electric lighting, phonographs, telegraphs, telephones, movies, and more. He also established one of the first modern research laboratories, which some scientists feel is the most important contribution of his vast legacy.

Born in 1847 in Milan, Ohio, Edison was the youngest of seven children. Although he was an industrious and clever child, it could well have been a serendipitous meeting that led Edison into a life of inventing. At the age of fifteen, Edison happened to be on hand to save the son of a telegraph operator from an oncoming railroad car. To thank him, the boy's father gave Edison free lessons in operating a telegraph. This led to a job with the Western Union Telegraph Company, which in turn awoke in Edison a lifelong interest in mechanics, electronics, and chemistry.

In 1868, Edison moved to Boston to continue working as a telegraph operator. There he displayed some of the first hints of his knack for innovation: He came up with a plan to improve a town fire-alarm system, and enhanced the printed outputs of telegraphs for the stockbrokers who relied on them for business. The next year, Edison moved to New York City, where he forged more friendships with prominent leaders in finance and industry. These connections would later prove pivotal in helping him launch the many business ventures tied to his inventions. For example, in 1870, Edison moved to Newark, New Jersey, and cofounded a stock ticker manufacturing company based on technology he helped develop. But it wasn't until 1876—when Edison established his famous Menlo Park laboratory—that the inventions started coming fast and furious.

Through his years of work with the telegraph, Edison had become an expert, and often fiddled with the new telephone technology, patented by Alexander Graham Bell in 1876, that allowed telegraphs to "speak." In one of his first significant inventions, Edison fine-tuned the telephone transmitter so that sound came through louder and clearer, making talking on the telephone easier and more practical. Next, in 1877, Edison unveiled his most spectacular invention to date: the phonograph. The ability to record and play back voices, sounds, and music seemed almost magical at the time, and Edison garnered fame and recognition around the world.

As if that weren't enough, the very next year, Edison came up with one of his brightest ideas yet: the electric light. Inventors around the world were already experimenting with the carbon arc light, which produced a bright, intense light for a short period of time. Edison bested them, though, by creating the incandescent lamp, which passed electricity through a wire filament, causing the filament to glow. This type of lighting was safer, longer-lasting, and more practical for home use than the carbon arc. In 1879, Edison and his associates figured out how to make a carbon filament out of burned sewing thread, and the precursor to the modern lightbulb was born. By 1880, Edison had perfected a 16-watt bulb that could burn for up to 1,500 hours. He also led the charge to create an industry to supply homes and businesses with the electricity necessary to power lightbulbs. As a result, he and other industry leaders banded together to form the General Electric Company in 1892.

Many other inventions and improvements upon existing technologies followed. Edison helped found the motion picture industry, established an iron ore processing plant, designed and improved batteries, and ventured into the cement business, among other forays. He was certainly one of the great scientific geniuses of our time, but he modestly refuted the notion that he was gifted and credited his success to hard work. As he was fond of saying, "Genius is one percent inspiration and ninety-nine percent perspiration."

An atomic explosion

Nuclear Reactor

Patent Name: Neutronic Reactor
Patent Number: 2,708,656
Patent Date: May 17, 1955
Inventors: Enrico Fermi, of Santa Fe, New Mexico, and Leo Szilard, of Chicago, Illinois

"I know not with what weapons World War III will be fought, but World War IV will be fought with sticks and stones." —Albert Einstein

What It Does

This device initiates and controls a self-sustaining nuclear chain reaction, and the massive amount of energy produced is typically used for power generation—but it has also been used as a weapon of mass destruction. The neutrons and fission by-products are used for various military, experimental, and medical purposes.

Background

On August 6, 1945, just three weeks after the July test in New Mexico, an atomic bomb was dropped on Hiroshima, Japan. Tens of thousands were instantly killed; 65 percent of the city's structures were damaged. Including radiation deaths within a year, the bombing killed more than 155,000 people and effectively ended World War II.

In 1938, German scientists Otto Hahn, Lise Meitner, and Fritz Strassmann discovered that by bombarding the nucleus of a uranium atom with thermal neutrons, the atom would split. In splitting one uranium atom, additional neutrons spontaneously formed that could then be used to split additional uranium atoms. The potential chain reaction was a mind-boggling discovery. As it splits into fragments, a heavy nucleus like the isotope of uranium can release several hundred million electron volts of energy. This discovery, a method to convert mass into energy, jump-started a race among nations to build the atomic bomb.

From 1939 to 1945, more than $2 billion was spent on what was then called the Manhattan Project. Directed by Robert Oppenheimer in Los Alamos, New Mexico, the project was a top-secret U.S. effort that involved scores of brilliant scientists, including Italian physicist and Nobel Prize recipient Enrico Fermi and his Hungarian colleague Leo Szilard, who worked out of a secret lab in Chicago. There were many challenges: The desired uranium-235 was difficult to extract, and it was nearly identical to uranium-238. As isotopes, their chemical compositions are similar, and ordinary extraction methods could not separate them efficiently. The scientists needed to develop a mechanical means to separate them, as well as a method for producing a self-sustaining nuclear chain reaction. Fermi and Szilard filed a patent for their invention in 1944, but it wasn't approved until 1955, six months after Fermi's death.

Although the original intention was to create weapons of mass destruction, today's nuclear reactors are viable power sources that provide 17 percent of the world's electricity. There are some four hundred nuclear power plants in the world, with about one hundred on U.S. soil.

Enrico Fermi

How It Works

The worst nuclear power accident occurred in 1986 in what is now Ukraine, in the former USSR. One of four reactors at the Chernobyl nuclear power plant lost control of a chain reaction, which caused explosions that blew the steel and concrete lid off of the reactor, unleashing massive amounts of radiation. More than 30 people died, and 135,000 people were evacuated.

A nuclear reactor enables a large but finite number of subsequent nuclear fissions to take place just below critical mass. Typically, such reactors are based on some variation of the following model:

1. Bundles of uranium rods that act as extremely high-energy heat sources are submerged in water.
2. Control rods that absorb neutrons are inserted into the uranium bundles.
3. Raising the control rods causes the uranium core to produce more heat; lowering them causes them to produce less.
4. The water becomes steam, which drives a turbine, which, in turn, spins a generator.
5. Housed inside a radiation-proof concrete liner is the reactor's pressure vessel, where the reactor core and other key operating components are located.

FIG.23.

Witnesses:

Inventors:
Enrico Fermi
Leo Szilard

By
Attorney.

In the Inventors' Words

"We have discovered certain essential principles required for the successful construction and operation of self-sustaining neutron chain reacting systems (known as neutronic reactors) with the production of power in the form of heat. These principles have been confirmed with the aid of measurements made in accordance with the means and method set forth in the above-identified application, and neutronic reactors have been constructed and operated at various power outputs, in accordance with these principles, as will be more fully brought out hereinafter."

CHRONOLOGY OF THE ATOMIC AGE

1896: Antoine Henri Becquerel conducted research on radiation, inspiring experiments two years later by Marie Curie.

1898: Marie Curie, a French physicist, working with her husband, Pierre, discovered the radioactive elements radium and polonium.

1905: Albert Einstein published three papers that astounded the scientific community and established three new branches of physics. One of the three papers confirmed the atomic theory of matter.

1911: Georg von Hevesy came up with the idea of using radioactive material as a "tracer." The principle was later applied to medical diagnostic procedures.

1938: German scientists Otto Hahn, Fritz Strassmann, and Lise Meitner demonstrated nuclear fission, showing the scientific community how a uranium atom can be split by bombarding it with neutrons. As the nucleus of the uranium atom is split, it loses mass, which is then converted into energy. The process can also start a chain reaction as other neutrons are formed, bombarding the nuclei of other atoms. The implications of this process on weapons development would soon be exploited.

Albert Einstein's letter to President Franklin D. Roosevelt explaining how to make an atomic bomb from a chain reaction of uranium

◀ **1939:** Albert Einstein wrote a letter to President Franklin D. Roosevelt, warning him that it was possible to build an extremely powerful bomb using atomic energy, and that Germany might already be in the process of doing so. Einstein's letter prompted the U.S. to begin an investigation into the development of the atomic bomb.

December 1941: Japan bombed Pearl Harbor, inciting the U.S. to enter World War II.

September 1942: The U.S. formed the Manhattan Project in a race to develop the atomic bomb before the Germans. Secret laboratories were soon established at various locations across the U.S.

November 1942: Robert Oppenheimer was named director of the atomic bomb laboratory in Los Alamos, New Mexico.

December 1942: Enrico Fermi and others at the University of Chicago achieved the first sustained nuclear chain reaction, heralding the dawn of the atomic age.

A photo of the Trinity test .25 seconds after detonation

◀ **July 1945:** Scientists from the Los Alamos laboratory detonated the first atomic device, which had a plutonium core and released as much energy as 18,600 tons of TNT. The test came to be known as the Trinity Test.

August 1945: The United States dropped two atomic bombs on Japan. On August 6, 1945, military forces dropped a uranium bomb on Hiroshima, killing 70,000 people and injuring as many. Three days later, the U.S. dropped a second atomic bomb, on Nagasaki, killing 40,000 and injuring 60,000. Japan surrendered, bringing an end to World War II.

0.025 SEC.
N

100 METERS

March 1946: Winston Churchill declared the existence of an "iron curtain" in Europe, created by the Soviet Union. By the end of the decade, U.S.-Soviet tensions escalated into what became known as the Cold War.

July 1946: The U.S. conducted nuclear bomb tests on the Bikini Atoll in the Pacific.

1949: The Soviet Union tested its first atomic bomb.

1950: President Harry S. Truman ordered the Atomic Energy Commission to investigate the possibility of building the H-bomb, or hydrogen bomb.

1951: Usable electricity was produced from nuclear energy at the National Reactor Testing Station in Arco, Idaho.

1954: The Atomic Energy Act of 1954 was passed, a measure that promotes the peaceful use of nuclear energy.

1957: The first full-scale nuclear power plant began operation in Shipping-port, Pennsylvania.

1961: President John F. Kennedy urged Americans to build fallout shelters in case of nuclear war.

1963: The U.S. and the Soviet Union signed the Limited Test Ban Treaty, which prohibited nuclear tests underwater, in the atmosphere, and in space. Underground nuclear tests were not banned.

1968: The Nuclear Nonproliferation Treaty was signed, which barred countries from making or assisting other countries in building nuclear weapons. Countries that didn't already have nuclear weapons pledged to refrain from building them. By 1986, 130 countries had ratified the treaty.

1978: The United States dropped plans to develop the neutron bomb, a specialized type of hydrogen bomb that destroys all life but leaves buildings intact.

1979: The Three Mile Island nuclear power plant near Harrisburg, Pennsylvania, suffered a partial meltdown. Low levels of radioactive material were released.

1983: The Nuclear Waste Policy Act of 1982 was signed, initiating the development of a nuclear waste repository.

1986: The Chernobyl nuclear reactor in USSR went through a meltdown and burst into flames. Over 135,000 people living within an 18-mile (30-kilometer) radius of the reactor were evacuated. Massive amounts of radioactive material were released into the environment.

1990: The Cold War formally ended with the Conference on Security and Cooperation in Europe.

1993: The Department of Energy turned its attention toward cleaning up nuclear contamination. Carefully controlled nuclear weapons arsenals still exist, but are considered deterrents to large-scale wars rather than signs of aggression.

2006 and beyond: North Korea has conducted successful nuclear tests and is developing their missile technology, which concerns the international community.

Penicillin

Patent Name: Method for Production of Penicillin
Patent Number: 2,442,141
Patent Date: May 25, 1948
Inventor: Andrew J. Moyer, assignor to the USA as represented by the secretary of agriculture, Peoria, Illinois

What It Does

Provides a practical, commercially viable way to mass-produce penicillin for use as an antibiotic.

Background

In the late 1950s, MIT professor John Sheehan was able to successfully synthesize penicillin. Sheehan received dozens of patents during his life. He worked on a variety of inventions, including a method of making explosives used in rockets and torpedo warheads.

Derived from penicillium mold, penicillin was discovered in 1928 by Sir Alexander Fleming, who observed that bread mold could kill colonies of bacteria. This observation, and the conclusions that Fleming drew from it, proved pivotal in the annals of medical history: Fleming had discovered that natural substances could be used to kill disease-causing bacteria in the human body. Penicillin would prove to be one of the most important antibiotics ever. The problem was in extracting the antibiotic from its source, the penicillium mold: It took too long to get too little.

Penicillium mold is a fungus that differs little from one that appears on bread in warm, humid weather.

During World War II, the need for penicillin became pressing, to treat a rising number of wounded. Struggling to find a way to produce it efficiently, scientists tried two general approaches. One approach was synthesizing the drug: Researchers worked to figure out the molecular structure of the antibiotic to duplicate it synthetically. The other approach was to speed up production of the natural antibiotic itself.

More than a thousand scientists attempted to uncover the mysterious molecular structure of penicillin to synthesize the molecule, to no avail. Meanwhile, international efforts concentrated on developing a means to mass-produce the natural antibiotic. Andrew Moyer, a microbiologist, led the research in the United States. In culturing the mold in lactose and corn steep liquor, Moyer discovered that he could successfully and dramatically expedite the production of the antibiotic. Although his technique was not patented until 1948, it saved thousands of lives during the war.

How It Works

The penicillium mold is cultured in lactose and corn steep liquor.

In the Inventor's Words

"This application is made under the act of March 3, 1883, as amended by the act of April 30, 1928, and the invention herein described, if patented, may be manufactured by or for the Government of the United States of

America for governmental purposes without the payment to me of any royalty thereon.

"Penicillin is a bactericidal or bacteriostatic compound that may be produced by *Penicillium notatum* Westling, as reported by Fleming (*Brit. Jour. Exper. Path.* 10,226 [1929]), who first observed the inhibition of bacterial growth in the vicinity of *P. notatum* colonies of petri-dish cultures. Penicillin possesses extremely valuable antibacterial properties which favor its application in the treatment of numerous infections. It is especially effective in the repressing or killing gram-positive staphylococci, pneumococci, gonococci, and numerous streptococci.

"I have found a method of increasing the penicillin content of culture liquors of a penicillin-producing mold, such as *Penicillium notatum* Westling, *Penicillium chrysogenum* Thom, *Penicillium baculatum* Westling and *Penicillium cyaneo-fulvum* Biourge, many-fold beyond that taught by the prior art, and to such an extent that production of penicillin for use as a therapeutic agent is now more feasible on a commercial basis. In addition to giving a better yield of penicillin, the present invention offers numerous other advantages, which will be apparent from the following description."

Refrigeration

Patent Name: Refrigerating Apparatus
Patent Number: 1,886,339
Patent Date: November 1, 1932
Inventor: Charles F. Kettering, of Dayton, Ohio

What It Does

Provides a safer means of refrigerating perishables, to keep food safe, edible, and appetizing for longer periods.

Background

Freon is a chlorofluorocarbon. CFCs were later discovered to cause damage to the ozone layer of our atmosphere, but are still in wide use today.

Refrigeration was around long before this patent was issued. Ancient cultures devised means of keeping meat in cold storage places; ice boxes were used in England in the early 1800s; technology led to the employment of gas compression to create heat-absorbing liquid gases. By the 1920s, some of the first electric gas-compressing refrigerators were already on the market. At the same time, leakage of some of the toxic gases being compressed as refrigerant was proving dangerous and in some instances fatal.

In 1928, industrial magnate and inventor Charles Franklin Kettering, then working for the Frigidaire Corporation, worked on developing a safer refrigerant. The result: Freon. It spawned a new method

An electric refrigerator from the 1930s

of refrigeration that has since been widely adopted as a standard. This patent represents the first commercially successful method of refrigeration to employ the new gas.

How It Works

1. Two generator-absorbers (10, 11) are charged with absorbent.
2. Absorbers connect by means of circuits to respective evaporator mechanisms.
3. When one generator-absorber is heated and the other is cooled simultaneously, liquid refrigerant is supplied to the evaporator.
4. The liquid is condensed, and gaseous refrigerant is withdrawn from the evaporator.

In the Inventor's Words

"The means for heating either generator-absorber is a burner (22) which may be placed at the lower extremity of the flue provided by the tube (15). The means for cooling either generator-absorber is a flue with which is associated a fan or blower (24) and which may be placed to direct a current of air through the flue (15). As illustrated, the blower is of the centrifugal type having its intake at the connection (27), but the particular form of air moving device is immaterial to the invention."

Rocket

Patent Name: Rocket Apparatus
Patent Number: 1,102,653
Patent Date: July 7, 1914
Inventor: Robert H. Goddard, of Worcester, Massachusetts

What It Does

This self-contained projectile propels itself into sustained flight through a series of combustions.

Background

Robert Goddard is the father of modern rocket science. While much of the world was still in awe of aircraft that could carry humans through the sky, Goddard was conducting experiments that would ultimately help carry humans far beyond our skies and into space.

Robert Goddard

This patent represents a first of its kind in illustrating the rocket propulsion theories for which Goddard became so famous. Just a year after he was awarded this patent, Goddard proved how thrust could be produced in a vacuum, bringing mankind one giant step closer to spaceflight. His expertise would first be applied to the military purposes at hand. During World War I, he developed solid-fuel rocket-launching weapons. This patent is for a rocket that uses solid fuels. He later developed liquid-fuel rockets that would be used to assist aircraft takeoff from naval carriers. He died in 1945, only a few years short of seeing the first rocket launched into space.

The apparatus described in this patent is a rocket specifically adapted to carry photographic devices. After the rocket finishes its flight, a second propelling charge sends the recording device into flight in an auxiliary rocket. Although not detailed in Goddard's patent, he mentioned that such devices could be safely returned to earth by means of a parachute.

How It Works

1. Ignition or detonation of a rocket sets off a series of combustions that serve to propel the apparatus into a sustained flight.
2. In figure 1, a cross-section of the rocket reveals a combustion chamber (12) with a tapered tube (11) at the base.
3. Within the tube, a fuse (14) is ignited.
4. A series of rotating disks within the combustion chamber is set into motion through consecutive explosions.

5. Figure 3 depicts one disk that includes around its periphery a filament (18) that ignites the explosive charges arranged on the disk.
6. When the propelling charge in the main rocket is sufficiently exhausted, a second fuse ignites the smaller rocket.
7. The auxiliary rocket is launched out of the firing tube (24).
8. Within the rocket head (29), a support is arranged to contain a recording instrument or camera.

In the Inventor's Words
"Mathematical analysis shows that in any rocket apparatus of given mass, the necessary propelling charge varies according to an expression in which the percentage of the heat energy of the charge which is transformed to kinetic energy enters in an exponential relationship. Hence, any increase in the efficiency in the transformation results in greatly increased velocity of the apparatus and also permits a reduction in the amount of explosives used."

Left: In 1892, building code changes allowed for steel cage construction, which made it possible to build the 22-story Flatiron Building.

Right: The Burj Khalifa in Dubai is currently the tallest building in the world.

Skyscraper Steel

Patent Name: Improvement in the Manufacture of Iron and Steel
Patent Number: 49,051
Patent Date: July 25, 1865
Inventor: Henry Bessemer, of London, United Kingdom

What It Does

Provides a strong material for construction of tall buildings and other massive structures.

Background

The word "skyscraper" was coined in the 1880s, long before some of the tallest structures were built. The tallest building in the world currently is the Burj Khalifa in Dubai, United Arab Emirates, made of glass, steel, and reinforced concrete, which stands at an amazing 2,717 feet, with 162 floors.

George Fuller built one of New York's first skyscrapers, the Flatiron Building, in 1902. Fuller understood how to build upward, by resolving load-bearing problems in structural design. He created steel caging within exterior walls, and he used a new type of steel, created by Englishman Henry Bessemer.

The sky became the limit when Bessemer developed a radical new iron- and steel-making process—the first such process for mass-producing steel inexpensively. Bessemer used a furnace to repeatedly blast air into a container of fluid metal, thereby removing carbon. The new process allowed for a significantly stronger material and offered a seemingly limitless dimension to the world of architecture.

This patent sets forth the principles of Bessemer's original process and introduces an improved use of large-scale metal molding casts. Bessemer included a crucial new concept in molding sturdier steel. Previous molds for pole-shaped structures, for example, included two plate casts that were sealed shut by lugs and bolts, as airtight as possible, to be removed after casting. Air would inevitably become trapped inside, weakening the material. Bessemer realized that leaving an opening created a natural vacuum, making the steel stronger. Whether Bessemer knew it at the time or not, his new methods would soon dramatically alter the skylines of cities around the globe.

How It Works

A chamber made of brick, firestone, or another slow conductor of heat features a raised outlet.

1. The chamber is filled to capacity with fluid metal; no leakage is allowed from the opening.

2. Air or steam is forced into the metal by means of a pipe, displacing the liquid and allowing for the escape of air through the outlet.
3. A tapping hole sealed with loam is punctured to allow the refined, decarbonized metal to flow into suitable molds.

In the Inventor's Words "My present invention relates, first, to peculiar modes by which the decarbonization or partial decarbonization and refinement of iron is effected by means of currents or jets of atmospheric air or steam (alone or mixed) which are made to impinge upon the surface or pass through or in contact with the metal while in a fluid state; secondly, in the manner in which the metal so treated is formed into ingots or masses suitable for being afterward made into bars, plates, or rods by the process of hammering or rolling."

Submarine

Patent Name: Submarine Vessel
Patent Number: 581,213
Patent Date: April 20, 1897
Inventor: Simon Lake, of Pleasantville, New Jersey

What It Does A submarine is an airtight and pressurized vessel for manned travel deep beneath the ocean's surface over a prolonged period. Submarines have been used in warfare as well as for various kinds of biological and oceanographic research.

Background Like many kids of his day, Simon Lake read the science-fiction adventure stories of Jules Verne, and his life's work would be directly influenced by the experience.

The United States Navy has four different classes of submarines: attack, which perform seek-and-destroy missions; ballistic missile, which provide strategic nuclear deterrence; guided missile, which complement the guided-missile platforms of cruisers and destroyers; and deep submergence rescue vehicles, which are used in response to a submarine accident.

After reading *Twenty Thousand Leagues Under the Sea*, Lake became determined to create a real-life *Nautilus*, a state-of-the-art submarine.

In the late 1800s, Lake and John Holland were the two most prominent figures in designing the modern submarine. Their ideas were different, however. Holland's submarines dove and surfaced at an angle; Lake's used control pumps to manipulate the boat's buoyancy, allowing it to move vertically in place. With wheels, Lake's submarine could rove along the sea bottom as well. Both designs had their appeal. While Holland did win a contract from the Navy, it was Lake who later solved the problem of vision by fashioning the earliest-known periscope. Lake is also credited with the development of basic submarine technologies including even-keel hydroplanes, ballast tanks, and a twin-hull design.

While still in his twenties, Lake built his first experimental submarine, naming it the *Argonaut, Jr.* Encouraged by its success, he formed the Lake Submarine Company of New Jersey the following year, and built another submarine: the *Argonaut.* In 1898, this submarine was considered the first to demonstrate successful operation and long-distance navigation in the open sea.

Lake's achievement inspired a letter from Jules Verne, who wrote: "While my book *Twenty Thousand Leagues under the Sea* is entirely a work of imagination, my conviction is that all I said in it will come to pass. A thousand mile voyage in the Baltimore submarine boat [the *Argonaut*] is evidence of this. This conspicuous success of submarine navigation in the United States will push on under-water navigation all over the world. If such a successful test had come a few months earlier it might have played a great part in the war just closed. The next great war may be largely a contest between submarine boats."

How It Works

In Figure 2:
1. Pumps are arranged to fill and empty tanks (i, j).
2. An engine (e) is powered by furnace (f).

The 36-foot-long *Argonaut I*, built in Baltimore at the Columbian Iron Works & Dry Dock, was powered by a 30-horsepower gasoline engine.

3. Weights (s) arranged below the vessel are connected by water-tight cables wound around casings (g).

4. Within the hollow keel (M), a shaft (c^1) is rotated by the engine, turning both the screw propeller (p) at the aft and the front wheel (B).

5. The submarine "driver" is stationed in a conning tower (I) with a steering wheel (K).

6. A companion hatch (O) is situated directly behind the conning tower.

7. A diver's room (M) is situated directly in front and below the conning tower.

8. Extending from a turret (H) behind the companion hatch by a ball-and-socket joint is an observation tube (G).

9. Individual crew members are hoisted inside the tube to obtain an unobstructed view of the horizon through a rounded heavy glass top (g^1).

In the Inventor's Words "My invention relates to an improved submarine vessel, and has for its object, first, to provide novel means for sinking the vessel to the bottom of the water when it is at a state of rest or has no headway and for permitting the vessel to rise to the surface of the water; second, to provide means whereby the vessel is enabled to travel upon the bottom or bed of the water; third, to provide a mechanism automatically controlled by the pressure of the water for submerging the vessel and maintaining it submerged at any desired or predetermined depth when under way; fourth, to provide means for automatically maintaining the vessel on a level keel irrespective of the disposal or shifting of the weights in the vessel; fifth, to provide novel means for affording ready ingress and egress from and to the vessel when submerged, and, lastly, to improve the construction generally and render more safe and certain the operation of submarine vessels."

Telegraph

Patent Name: Improvement in the Mode of Communicating
Information by Signals by the Application of Electro-Magnetism
Patent Number: 1,647
Patent Date: June 20, 1840
Inventor: Samuel F. B. Morse of New York, New York

"What hath God wrought?" —Samuel Morse's first telegraph message

What It Does

Offers a system to relay audible messages by means of code over long distances via electromagnetic waves.

Background

Western Union was established in 1851 and, in a decade's time, had built the first transcontinental telegraph line. The first transatlantic cable offered service in 1868.

Samuel F. B. Morse

A different telegraph system using the same principles of electromagnetism was patented in Great Britain in 1837.

If Samuel Finley Breese Morse was skeptical about the nature of technological advancement, his life belied the fact. He abandoned a successful career as a portrait painter to embark upon the improvement and creation of technology as an inventor. In the 1830s, he became interested in electronics and telegraphy. He envisioned a system of relaying messages in an electronic alphabet of dots and dashes through electromagnetic waves. His first successful system involved a tape on which a register printed his self-named Morse code language of dots and dashes. A telegraph operator would move a key on the device that would manipulate the circuit of electricity to transmit a series of electric pulses. The longer pulses were expressed as dashes, the shorter ones dots—and each letter of the alphabet was translated into a few specific dots and/or dashes, forming the code that was transmitted by the telegrapher. For example, the letter S was expressed as three dots, or short pulses; the letter O as three dashes, or long pulses; so the message "SOS," used as a distress signal, would be three dots, three dashes, then three dots again.

A line was constructed from Washington to Baltimore, and Morse relayed his famous first message in 1844 from inside the Capitol Building. Within a decade, thousands of miles of wire stretched throughout the country, connecting major cities, enabling a more rapid expansion of the American West, and revolutionizing long-distance business transactions. His invention is regarded as the earliest precursor to the "information highway" or Internet. Morse's first telegraph message was ominous, and his question remains unanswered—hidden, perhaps, deeper now in the human imagination of the computer age.

How It Works

1. A circuit of electric conductors contains one or more electromagnets.
2. At the message-originating end of the circuit, an operator manipulates a signal lever that breaks and reconnects the circuit of conductors.
3. Electric pulses transmit along the circuit.

4. At the message-receiving end, a register prints out a coded language of dots and dashes derived from the short or long pulses.

5. Deciphering the code of dots and dashes reveals the messages.

In the Inventor's Words

"Be it known that I, the undersigned, SAMUEL F. B. MORSE, of the city, county, and State of New York, have invented a new and useful machine and system of signs for transmitting intelligence between distant points by the means of a new application and effect of electro-magnetism in producing sounds and signs, or either and also for recording permanently by the same means, and application, and effect of electromagnetism, any signs thus produced and representing intelligence, transmitted as before named between distant points; and I denominate said invention the 'American Electro Magnetic Telegraph'…"

An early telephone operator

Telephone

Patent Name: Improvement in Telegraphy
Patent Number: 174,465
Patent Date: March 7, 1876
Inventor: Alexander Graham Bell of Salem, Massachusetts

What It Does

Enables people to contact one another and communicate verbally over long distances.

Background

Bell also invented the photophone, which transmitted sound on a beam of light. The invention was a direct precursor to the fiber optics used in telecommunications today. But because Bell was unable to resolve problems associated with light source interference, the revolutionary principles of the photophone were only much later appreciated.

When Alexander Graham Bell died on August 2, 1922, telephones in the United States stopped ringing. Citizens shared a minute of silence to honor the inventor.

At the age of twenty-nine, Alexander Bell was awarded this patent for an improved system of telegraphy that allowed multiple messages to be relayed simultaneously through varying harmonic vibrations. Instead of a signal lever sending interruptions over the wire, the human voice itself could send the signal. A year later, Bell established the Bell Telephone Company. His invention marked the end of the telegraph system as it was then known and the beginning of the telephone.

Sound had been a constant interest throughout Bell's life. He was born in Scotland in 1847 to a mother who was almost completely deaf and a father who was involved in speech pathology. Like his father, Bell became devoted to the deaf, teaching speech techniques at a school in Boston beginning in 1871. Bell was also a firm believer that science and technology could improve human life, and he possessed a natural knack for invention.

Already quite familiar with the various vibrating pitches of the human voice, he began work on what he called a "harmonic telegraph." Bell maintained that a system based on a number of transmitting instruments with different rates of vibration could simultaneously make and break an electric circuit to transmit a fluid, continuous current over a sound-carrying wire.

Bell met Thomas Watson, a man highly regarded for his abilities improving machines and helping other inventors. Watson would soon help Bell improve on the finer points of his invention. On March 10, 1876, Bell spoke the first words into his machine, beckoning his assistant: "Mr. Watson, come here. I want you!" In another room, Watson had received the world's first telephone call.

How It Works

1. A speaker speaks through vibrating membranes that play on the electromagnetic currents to relay an electronic facsimile of his voice over the wire.
2. Figure 7 of the patent illustration depicts an armature (c) loosely fastened to a leg (d) of an electromagnet (b).
3. The other end attaches to a stretched membrane (a).
4. A cone (A) converges sound vibrations through the membrane.
5. The vibration of the membrane enacts the motion of the armature, creating undulations upon and through the circuit (E, b, e, f, g).
6. The receiving end duplicates the vibrations so that a similar sound can be heard at another cone (L).

In the Inventor's Words "My present invention consists in the employment of a vibratory or undulatory current of electricity in contradistinction to a merely intermittent or pulsatory current, and of a method of, and apparatus for, producing electrical undulations upon the line-wire."

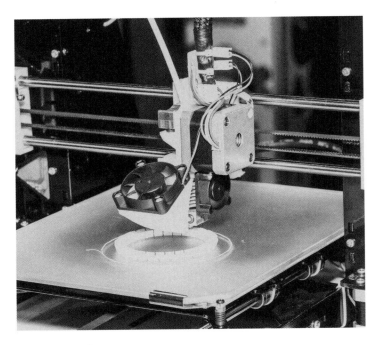

A 3D printer

3D Printer

Patent Name: Apparatus for Production of Three-Dimensional Objects by Stereolithography
Patent Number: 4575330 A
Patent Date: March 11, 1986
Inventor: Charles W. Hull, of Valencia, California

What It Does This machine creates three-dimensional models or structures by depositing the material layer by layer.

Background The first patent for a 3D-printing device was issued to Charles Hull in 1986, the same year Hull started the company 3D Systems, though the full scope of the technology's applications wasn't widely realized until the twenty-first century. 3D printing, also called additive manufacturing, uses a variety of technologies and materials, but these variations all have their roots in a process called stereolithography. Models, structures, and patterns are constructed out of synthetic liquid plastics, deposited in layers by an automated machine with a nozzle.

The underlying technology used in stereolithography is photopolymerization. Polymerization is the process by which molecules are connected to form larger molecule chains or three-dimensional molecular structures. Photopolymerization uses light to stimulate this process, generally ultraviolet (UV) light because of its high energy level. A typical 3D printer uses an acrylic-based material that remains liquid until it is hit with UV light, causing the substance to rapidly turn solid.

The technology has evolved and expanded significantly in recent years, however. Some 3D-printing techniques weld powdered materials together to form objects, such as direct metal laser sintering. DMLS uses a high-powered laser, often a fiber laser with ytterbium optical fibers, to *sinter*, or fuse together, a bed of metal powder into a solid object. Additional layers of metal powder can then be added and microwelded with the laser, ultimately creating a complete metal object without requiring any tooling. The speed and automation of DMLS make it a popular manufacturing method in the aerospace industry. SpaceX, for example, uses direct metal laser sintering to construct its SuperDraco rocket engines out of Inconel, a nickel- and chromium-based superalloy, for use on its Dragon spacecraft.

General Electric has invested more in large-scale manufacturing using 3D printing than any other company. Its Center for Additive Technology Advancement opened near Pittsburgh, Pennsylvania, in September 2016 to 3D-print metal components. And a series of "brilliant factories" from GE combine 3D-printing technologies with robotic automation and laser scanning; the first opened in Pune, India, in 2015. Economic theorist Jeremy Rifkin agues that 3D printing is such a major breakthrough in digital manufacturing that it will usher in a new industrial revolution, akin to the assembly line's impact on manufacturing in the early twentieth century.

How It Works

1. A UV curable liquid plastic, or something similar, is deposited on a moving surface, controlled by a computer, to create a base for the object.

2. A nozzle deposits the liquid material, layer by layer, on the base while the surface moves laterally and horizontally to align the nozzle with the desired area. The nozzle and the movement of the platform are controlled by a computer control system.

3. A UV light source, also controlled by the computer, targets the freshly deposited material, causing it to cure and harden.

4. Mirrors and other optical devices can be used to spread the UV light over a wider area, further curing the material.

5. The device continues to layer on material in this way to create a three-dimensional object.

6. Alternatively, metal powders can be superheated with a laser to weld the powders into solid layers, which can be stacked and sintered together to form an entire object.

Fig. 5

COLUMNATED, BROAD
ULTRAVIOLET
LIGHT SOURCE

35

23   36

21   22   30   30a   30b   30c

29

Fig. 6

38

23   39

21   40   30a   30b   30c

22   30

29

In the Inventor's Words "It's really blossomed just in the last few years—in the sense of really rapid growth and recognition. There's a lots of things that contributed to that, I think: a lot of the medical applications catch people's imagination; certainly the maker movement, with low-cost machines getting hobbyists interested in inventing and building using 3D printing."

Third-Generation Wireless Mobile Telecommunications (3G)

Patent Name: Mobile Internet Access
Patent Number: US 6618592 B1
Patent Date: September 9, 2003
Inventors: Harri Tapani Vilander and Tom Mikael Nordman, of Stockholm, Sweden

What It Does

This system allocates Internet Protocol (IP) addresses to mobile devices using a mobile telecommunications network, allowing users to browse the Internet on a wide variety of mobile devices, such as cell phones.

Background

A fundamental transformation happened to mobile phones around the turn of the millennium. The first- and second-generation mobile telecommunications networks, often called 1G and 2G, allowed for analog cellular phones first, like those commonly used in the 1980s, and digital cellular phones second, such as the cell phones and flip phones used in the 1990s and early 2000s. Analog and digital cell phones both use radio technology to transmit voice signals, but digital signals are converted to a binary sequence of 0s and 1s, allowing them to be compressed so that more signals can be sent within a specific bandwidth.

3G allows people to access the Internet on a wide range of mobile devices, like this iPad.

The Universal Mobile Telecommunications System (UMTS) was an early third-generation, or 3G, mobile telecommunications network. It established widespread "mobile broadband" coverage, allowing access to the Internet through a Web browser, as well as improved voice calling, video calling, and mobile streaming. 3G made several improvements to existing networks, including data-link systems that increased peak data-transfer speeds significantly. Connection times were reduced, and the networks were expanded to support a higher number of devices. Although the first phones to access the Internet used 2G, it wasn't until the widespread implementation of 3G technology in the 2000s that manufacturers like BlackBerry and Apple started developing devices specifically to access the mobile Web.

The BlackBerry Quark series of phones, released in 2003, were among the very first cell phones to include an Internet browser. Although cell phones with Internet access were primarily used by affluent individuals at first, these phones gradually became more widespread, particularly after the release of the iPhone in 2007. Additional technologies soon became common in mobile devices, such as GPS mapping and video streaming, which were made possible by 3G.

Since 1G systems were introduced in the 1980s, there has been a new generation of mobile telecommunications technologies approximately once every decade, with 2G networks launched in 1991 and the first commercial 3G technologies established in the United States in 2003. Today, new mobile devices use 4G networks, which must meet a set of requirements established by the International Telecommunication Union Radiocommunication Sector (ITU-R) in 2008. 4G networks, such as LTE, offer higher data-transfer rates, though 4G is more of an evolution of 3G than a radical transformation of the technologies used, as was seen in the shift from 2G to 3G.

How It Works

1. The Global System for Mobile Communications (GSM) establishes a set of standards for mobile telecommunications services to advertise as 3G, or third generation.

2. 3G services are based on the existing wireless network, which is made up of base stations that use transceivers to transmit and receive signals from multiple sources, such as phones and cellular antenna towers, across a given region.

3. Signals are passed from cell tower to cell tower until the tower closest to the mobile device establishes a connection with it.

4. Transferring data through these connections allows a mobile device to make phone calls, access the Internet, use video calling, download audio or video, and use mobile streaming services, among other capabilities.

5. While mobile phones are the primary devices that use 3G and 4G networks, other electronics can access mobile broadband networks as well, such as laptop computers and dongles that plug into USB ports.

In the Inventors' Words

"It is an object of the present invention to overcome or at least mitigate the above noted disadvantages of existing and currently proposed mobile telecommunications networks. In particular, it is an object of the present invention to eliminate or at least obviate the need to negotiate an IP address between a mobile terminal and [a] node of the mobile network prior to commencing an Internet session."

Figure 3   Figure 4

A vintage Underwood typewriter

Typewriter

Patent Name: Improvement in Type-Writing Machines
Patent Number: 79,265
Patent Date: June 23, 1868
Inventor: C. Latham Sholes, Carlos Glidden, and Samuel W. Soule, of Milwaukee, Wisconsin

What It Does

Prints uniform characters onto paper by means of a system of push buttons that activate metal dies that impress an ink-infused ribbon onto the paper.

Background

The "QWERTY" keyboard is named after the first six letters in the row second from the top. Its design is the result of a shrewd observation: Letters placed close to each other that are commonly used together had a tendency of jamming the keys during typing. Sholes sold his patent rights for $12,000 to the rifle manufacturing company Remington & Sons in 1874, which made and marketed the machines.

American humorist Mark Twain purchased one of the early model Remington typewriters and is believed to be the first author to submit a manuscript in type to a publisher.

In the 1400s, a German goldsmith named Johannes Gutenberg invented a printing press with replaceable letters that made printed materials available to the masses. It remained the standard until the twentieth century. Not until some 450 years later did a successful personal printing press of sorts—a Type-Writing Machine—make its appearance using similar principles of moveable type. It followed hundreds of previous attempts to produce similar instruments, dating back to the early 1700s. In fact, this patent is a precursor to the one many consider to be the first typewriter.

Sholes was a newspaper man. He served as editor for the *Wisconsin Enquirer* in Madison, the *Milwaukee News*, and the *Milwaukee Sentinel.* In 1864, he and his friend Samuel Soule patented a page-numbering machine. Carlos Glidden, a fellow inventor, suggested that the device might be reworked into a letter-printing machine.

Glidden, Soule, and Sholes were granted a patent on June 23, 1868, for a typewriter that was an improvement of a prior invention for which they had applied for a patent. A few years later, Sholes and friends developed an even better version, which would ultimately become known as the first successful commercial typewriter. The later version established the modern "QWERTY" keyboard, which remains the standard keyboard design today.

How It Works

1. When pressed, the lettered and numbered keys activate hammers with individual characters stamped on them, which in turn strike an inking ribbon against a piece of paper.
2. Each stroke engages the motion of the inking ribbon, which winds from one spool to another.
3. Also, by means of a cord and pulley system, keystrokes cause the carriage on which the paper is set to move along a set of ratchets, advancing it at each stroke until it reaches the end of a line and must be returned to the next line manually, by means of a lever.

In the Inventors' Words "Be it known that we, C. Latham Sholes, Carlos Glidden, and Samuel W. Soule, of the city of Milwaukee, and county of Milwaukee and State of Wisconsin, have invented new and useful Improvements in Type-Writing Machines; and we do hereby declare that the following is a full, clear, and exact description of the invention, which will enable those skilled in the art to make and use the same, reference being made to the accompanying drawings, forming part of this specification.

"This invention is of improvements to an invention of a type-writing machine, an application for a patent for which we filed October 11, 1867. Its features are a better way of working the type-bars, of holding the paper on the carriage, of moving and regulating the movement of the carriage, of holding, applying, and moving the inking-ribbon, a self-adjusting platen, and a rest or cushion for the type-bars to follow.

"Thus made, the type-writer is the simplest, most perfectly adapted to its work—the writing of ordinary communications with types instead of a pen—and in every way the best of all machines yet designed for the purpose, particularly as to the cost of making the machine and the neatness and labor-saving quality of its work."

A worker cracks crude rubber at the Goodyear Tire factory in Akron, Ohio, in 1928.

Vulcanization of Rubber

Patent Name: Improvement in India-Rubber Fabrics
Patent Number: 3,633
Patent Date: June 15, 1844
Inventor: Charles Goodyear, of New York, New York

What It Does   Provides a method of producing a durable rubber material, especially for the manufacture of auto and other tires.

Background   Sometimes a person's contributions are most celebrated after he or she is dead. Charles Goodyear's life followed this pattern. While experimenting with some raw rubber he had acquired from a shoe company, the poor hardware merchant was thrown in debtor's prison; he couldn't quite afford his investment. When he was released in 1839, he continued experimenting with rubber, adding ingredients to his recipe. He eventually produced a gummy substance in the shape of a ball, which was accidentally tossed onto a hot

Vulcanized rubber met many emerging needs of the day, including a practical sole for sneakers and pneumatic tires with inner tubes for bicycles, developed by John Dunlop in 1888.

The first "all-weather, all-season" tire for passenger cars was manufactured in 1927 by the Goodyear Tire Company.

According to Goodyear, billions of rubber trees are now cultivated; some three million tree "milkers" harvest the crop; and the United States imports almost half of it.

stove. The inadvertent process of vulcanization had begun. As the material melted, Goodyear identified the missing ingredient: heat. His material eventually hardened to a consistency he'd been trying to achieve from the beginning.

This proved to be a monumental accomplishment that has since altered the ways of the world: A pliable, sturdy rubber-based material could be molded into varying shapes and applied to any number of uses, the most notable of which is the production of tires. Sadly, Goodyear's invention did not garner him instant fame and fortune. A company named in his honor began making synthetic tires nearly a century later, in 1937.

As Goodyear's corporate history relates it, the hot stove incident was more than a happy accident; it was a tribute to Charles Goodyear's creativity. "It held meaning for the man whose mind was prepared to draw an inference," and who had "applied himself most perseveringly to the subject." Goodyear himself wrote: "Life should not be estimated exclusively by the standard of dollars and cents. I am not disposed to complain that I have planted and others have gathered the fruits. A man has cause for regret only when he sows and no one reaps." Charles Goodyear was a poor man when he died, but many have achieved great wealth as a result of his innovations.

How It Works

A mixture of twenty-five parts India rubber, five parts sulphur, and seven parts white lead is heated to a temperature between 212° and 350° Fahrenheit, making a durable, flexible substance suitable for such high-impact products as car tires.

In the Inventor's Words

"My principle improvement consists in the combining of sulphur and white lead with the India-rubber, and in the submitting of the compound thus formed to the action of heat at a regulated temperature, by which combination and exposure to heat it will be so far altered in its qualities as not to become softened by the action of the solar ray or of artificial heat at a temperature below that to which it was submitted in its preparation—say to a heat of 270° of Fahrenheit's scale—nor will it be injuriously affected by exposure to cold."

Vulcanization allows rubber to be waterproof and cut easily into shapes like rain boots.

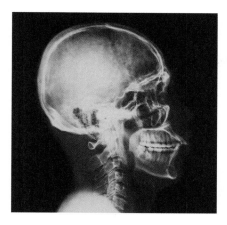

The widespread availability of X-rays revolutionized health care.

X-Ray Tube

Patent Name: Vacuum Tube
Patent Number: 1,203,495
Patent Date: October 31, 1916
Inventor: William D. Coolidge of Schenectady, New York, assignor to General Electric

What It Does

Offers an efficient means of producing X-rays in order to capture images of the interiors of opaque bodies.

Background

For his discovery of X-rays, Wilhelm Conrad Röntgen received the first Nobel Prize in Physics, in 1901. In his honor, the unit of radiation exposure equal to the quantity of ionizing radiation that will produce one electrostatic unit of electricity in one cubic centimeter of dry air at 0°C and standard atmospheric pressure is called a "röntgen."

Coolidge was awarded eighty-three patents in his life, ranging from radar systems to comfort devices such as the electric blanket. He was inducted into the Inventors Hall of Fame in 1975.

X-rays are electromagnetic waves, and they are very penetrative, as Wilhelm Conrad Röntgen inadvertently discovered in 1895. That fateful day, the German physicist was experimenting with a cathode-ray tube, an apparatus that displays images by means of a scanned electron beam. But Röntgen observed the effects of rays that extended beyond the range of cathode rays—rays that traveled through wood and aluminum. The rays were being generated where the electron beam hit against the inside of the tube. Because he was unfamiliar with them at the time, he called his discovery "X-radiation." His associates called it "Röntgen-radiation" or "Röntgen" rays.

Some of the benefits these X-rays offered were instantly apparent: They would provide a unique glimpse into the human body. Tissues, muscle, bones, and fat vary in their absorption of X-rays. Because they made distinct pictures of the body's bones and organs, X-rays revolutionized radiology and medicine.

Subsequent efforts to develop a practical method of generating X-radiation led American electrical engineer William D. Coolidge to invent his now-famous X-ray tube. Prior to this, Coolidge, who was working for General Electric, invented ductile tungsten to be used as a refined filament, now standard in electric lighting.

Previously, X-rays had been produced by bombarding an enclosed space with a steady flow of electrons, which, upon colliding, underwent a sudden change in velocity and were transformed into penetrative rays. In his research with X-ray tubes, Coolidge found many problems with the existing ways of generating the radiation. In previous tubes, for example, the pressure of the gas used to affect the flow of electrons was too variable, resulting in a variable and inconsistent penetrative power of X-rays.

Coolidge decided to remove the gas used to facilitate bombardment of electrons and, instead, heat a tungsten filament cathode to incandescence. He found that this would emit electrons that could be manipulated into the desired effect by electromotive force, while eliminating the variability of gas pressure.

How It Works

1. Figure 1: A glass housing (1) surrounds a plate-shaped tungsten anode (3) extended into the center by a tungsten stem (4) from a tubular end (2).
2. Lead-in wires (6, 7) are adapted to a power supply and sealed into the glass support, where their energy heats cathode (5).
3. Surrounding the cathode is a hollow cylinder of tungsten (11), which acts as a static modifier to steady the focus of the cathode discharge.

In the Inventor's Words

"My invention relates to vacuum tubes and more especially to tubes operated for the purpose of producing Röntgen or X-rays. The tube which I have produced differs so radically from the tubes of the prior art used for producing Röntgen or X-rays as to amount not so much to an improvement on prior tubes as to an entirely new variety of tube differing both in its principles of operation and in its operating character."

Two

The Little Patent That Could

Aerosol Can

Patent Name: Method and Means for the Atomizing or Distribution of Liquid or Semi-Liquid Materials
Patent Number: 1,800,156
Patent Date: April 7, 1931
Inventor: Erik Rotheim, of Oslo, Norway

What It Does

Dispenses or distributes liquid or semiliquid material in "atomized" spray form.

Background

Erik Rotheim invented the forerunner of the can-and-aerosol system we now know as the aerosol spray can. He received a Norwegian patent on October 8, 1926, nearly five years before he was granted a U.S. patent for his invention. The simple practicality of this little piece of ingenuity has made it ubiquitous: Take a look in your own home and you'll surely find such necessities as spray paint, spray deodorant, hairspray, aerosol cleaning products, and lots more.

During World War II, U.S. Department of Agriculture researchers came up with a device pressurized by a liquefied gas for use as a compact bug repellent. It closely resembled today's aerosol can. Further modifications reduced or eliminated the release of fluorocarbons, which were found to damage the ozone layer.

Where there's a will, there's a way, and apparently many people have a will to get high. Aerosol inhalant abuse is a serious matter, with some worrisome statistics suggesting that around 7% of eighth graders use inhalants.

How It Works

1. A short pipe (7) at the bottom of a container (1) serves to fill the bottle while under pressure, being closed by compression and soldered shut when filled.
2. Near the top, a valve (4) is held in place by the action of the inner pressure on the spring (8).
3. Connected to the valve is an opening (13) in the ascending pipe (3).
4. A lockable pressure knob (5) and an ejector pipe (6) serve as both a stop and a spraying orifice.

April 7, 1931.
E. ROTHEIM
1,800,156
METHOD AND MEANS FOR THE ATOMIZING OR DISTRIBUTION
OF LIQUID OR SEMILIQUID MATERIALS
Filed Sept. 30, 1927

In the Inventor's Words

"When an outlet opening in the container enclosing the material and the condensed dimethyl ether is opened the material will be forced out under the pressure prevailing in the container. By suitably constructing the outlet it is possible to cause the material to pass out in the form of a solid jet or as an…atomized spray."

Aspirin

Patent Name: Acetyl Salicylic Acid
Patent Number: 644,077
Patent Date: February 27, 1900
Inventor: Felix Hoffman, of Elberfeld, Germany

What It Does   Relieves pain associated with headache, inflammation, and fever.

Background   In the early 1800s, scientists had been able to isolate salicin, a compound found in willow plants, which was known to provide pain relief when

Felix Hoffman

ingested. A series of extraction methods followed, but one common complaint about ingesting salicin remained: It irritated the stomach. In 1853, a French chemist named Charles Frederic Gerhardt added sodium and chloride to alleviate the symptom, but he neglected to market it and the formula became lost.

Felix Hoffman, a German chemist working for the Bayer Company, came across the work on acetyl salicylic acid several decades later. In 1899, the compound formula was marketed by Bayer under the trademark Aspirin®. It was sold in powder form until the first tablets came on the market in 1915. Bayer tried to maintain sole rights to the manufacture of the drug but was forced to relinquish its trademarks in 1919 following the Treaty of Versailles.

How It Works

Aspirin belongs to a class of drugs called salicylates. It is the active ingredient in more than fifty over-the-counter medications. Aspirin inhibits an enzyme that triggers many biological processes, such as tissue inflammation.

In the Inventor's Words

Felix Hoffman was seeking a way to ease his father's arthritis, when he invented acetyl salicylic acid.

If it was ever published in its time, Hoffman's patent application must have read like a slap in the face of a pharmaceutical rival: "I have now found that on heating salicylic acid with acetic anhydride a body is obtained, the properties of which are perfectly different from those of the body described by Kraut. According to my research the body obtained by means of my new process is undoubtedly the real acetyl salicylic acid… Therefore the compound described by Kraut cannot be the real acetyl salicylic acid, but is another compound."

VIEW OF THE WORKS OF
J.L.ELLWOOD & CO.
De Kalb, Ill.
SOLE MANUFACTURERS OF THE
GLIDDEN STEEL BARB FENCE WIRE
FOR THE
WESTERN STATES & TERRIT'S

VIEW OF THE WORKS OF
WASHBURN & MOEN MANUFACTURING CO
Worcester, Mass
SOLE MANUFACTURERS OF THE
GLIDDEN STEEL BARB FENCE WIRE
FOR THE
EASTERN & SOUTHERN STATES

FOR SALE BY

Joseph Glidden

Barbed Wire

Patent Name: Improvement in Wire-Fences
Patent Number: 157,124
Patent Date: November 24, 1874
Inventor: Joseph F. Glidden, of De Kalb, Illinois

What It Does

This wire features sharp protrusions or "barbs." Fences made from it are an effective deterrent to whatever (or whoever) is being kept in or out.

Background

Barbed wire is not just used to contain livestock. It also secures industrial property, helps keep prisoners from escaping, safeguards temporary stations during wartime, and contains prisoners of war.

As native people of North America witnessed the open ranges being increasingly staked, sectioned off, and fenced in, they coined a nickname for barbed wire: the devil's rope.

The invention of barbed wire represented a beginning and an end in American history. It marked the end of the free-ranging cowboy, the roving pioneer, and the nomadic Native American tribes; it marked the beginning of a new industry based on contained cattle-ranching operations and it further opened the West to settlers. Those who favored the open range vehemently opposed the improved fencing technology. Cutting barbed wire was a felony.

Joseph Glidden was born in Clarendon, New York, and moved to De Kalb, Illinois, in his late twenties. By 1873, other people were already working on ways to improve the existing simple wire fencing. It was too easily broken by the weight of a heavy heifer. Attempts were made to devise a pain-inducing wire that would teach cattle to "steer" clear. The sixty-year-old Glidden invented a form of fencing that featured barbs placed at intervals along one wire, with another wire twisted around the first wire, holding the barbs in place. He received a patent for the "barbed wire" that same year and sold half of his patent rights to Isaac Ellwood. Together, they built a factory in downtown De Kalb to manufacture the product.

How It Works

1. Barbs are set in place along a wire.

2. This wire and a companion wire are stretched between a hook at one fence post and a thumbpiece securely fastened to another.

3. The thumbpiece is designed to transversely twist the wires together, thereby securing the barbs in place.

In the Inventor's Words

"This invention has relation to means for preventing cattle from breaking through wire-fences; and it consists in combining, with the twisted fence-wires, a short transverse wire, coiled or bent at its central portion about one of the wire strands of the twist, with its free ends projecting in opposite directions, the other wire strand serving to bind the spur-wire firmly to its place, and in position, with its spur ends perpendicular to the direction of the fence-wire, lateral movement, as well as vibration, being prevented. It also consists in the construction and novel arrangement, in connection with such a twisted fence-wire, and its spur-wires, connected and arranged as above described, of a twisting-key or head-piece passing through the fence-post, carrying the ends of the fence-wires, and serving, when the spurs become loose, to tighten the twist of the wires, and thus render them rigid and firm in position."

Bar Code

Patent Name: Classifying Apparatus and Method
Patent Number: 2,612,994
Patent Date: October 7, 1952
Inventor: Norman J. Woodland, of Ventnor, New Jersey, and Bernard Silver, of Philadelphia, Pennsylvania

What It Does

Scans merchandise for classification and pricing purposes.

Background

On June 26, 1974, the first product with a bar code was scanned at a checkout counter in Troy, Ohio. It was a ten-pack of Wrigley's Juicy Fruit Chewing Gum. You can see it displayed at the Smithsonian Institution's National Museum of American History.

Next time you're stuck in line at the grocery store behind someone who's paying with pennies and outdated coupons, count your blessings. At least you're not waiting for the cashier to enter each item's price manually into the cash register. That was a fact of life when you shopped before the bar code and the bar code scanner came into being.

The bar code system is a time-saving technology that has significantly changed the consumer purchasing experience wherever it has been employed. In 1949, a local grocer asked two Drexel Institute of Technology grad students to automate product checkout. Eventually, Norman Woodland and Bernard Silver devised a system that employed a coded pattern and photosensitive technology that could read the codes. Their system is quite similar to the one used today. Their bar code had four lines. The first was a reference line, and the other three specified information relating to the first line. Using more lines allowed more specific and numerous classifications to be coded. But it would be a while before industry standards were set. The bar code's first practical use was matching railroads with their correct trains.

How It Works

1. A bar code displays relevant information through classifiable patterns of reflected and nonreflected light.

2. In Figure 11, the positions and presence of white lines against a black background represent the datum line, or bar code (12), on a product (11).

3. The highly reflective white lines move facedown against a transparent conveyor (13), where they are subject to multiple sources of illumination (14).

4. The reflected emanations are immediately projected into an photosensitive optical and scanning element (15), which includes a sensing mechanism that moves back and forth across the conveyor.

5. Later innovations in bar code technology have employed holographic image-scanning.

In the Inventors' Words "This invention relates to the art of article classification and has particular relation to classification through the medium of identifying patterns.

"It is an object of the invention to provide automatic apparatus for classifying things according to photo-response to lines and/or colors which constitute classification instructions and which have been attached to, imprinted upon or caused to represent the things being classified."

A Galvanic battery

Battery

Patent Name: Galvanic Battery
Patent Number: 373,064
Patent Date: November 15, 1887
Inventor: Carl Gassner, Jr., of Mentz, Germany

What It Does

Provides a portable, wireless energy source for electronically powered devices.

Background

The battery is a little invention that just keeps on ticking. The recent rise in cordless products, including cell phones, video cameras, tools, and toys, keeps the battery a busy component in modern life. More than three billion batteries are purchased each year in the United States.

Alessandro Volta "powered" a new field of invention when he created the "Voltaic pile." Others before him, including Benjamin Franklin, had tried harnessing electricity. Volta devised the basic model: the conduction of electricity between two opposite charges. It is the basis on which later improvements to batteries would build. Subsequent innovations involved connecting cathodes with liquid electrodes. They were effective, but also heavy and easily breakable.

One of the most significant modifications to the battery was developed by Carl Gassner, who invented the "Galvanic battery"—the first commercially successful "dry" cell battery. Gassner used a zinc container and a porous material to absorb the electrolyte, and sealed the cell across the top with bitumen. Gassner's invention instantly made the battery more practical and easier to handle. Additionally, Gassner's addition of zinc chloride helped minimize the zinc corrosion, extending the natural life of the battery. It is the forerunner of today's ubiquitous carbon-zinc batteries.

How It Works

When a negative electrode is put in contact with a positive electrode, it causes a chemical reaction, which produces electricity. In the battery, the electrodes are divided by an ion-conducting separator and housed within an ion-conducting electrolyte. The reaction is triggered when the cell is connected to an external load, conducting a current of electrons from the negative electrode back to the positive electrode. Gassner's battery works as follows:

1. An exciting agent—the electrolyte—contains by weight the following: one part oxide of zinc; one part sal ammoniac; three parts plaster; one part chloride of zinc; and two parts water.
2. A zinc cylinder—the container—houses an isolated cylinder of carbon manganese.
3. The space between the two cylinders is filled with the exciting agent, in liquid or semiliquid form, which becomes relatively solidified.

In the Inventor's Words

Volta's "Voltaic Pile" alternates between silver (or copper) and zinc disks, which together produce an electric current.

"The oxide of zinc may be employed with great advantage as a constituent of any well-known exciting agent for the elements, with which it can be mechanically mixed and introduced into the galvanic cell and act therein, as hereinafter set forth. I do not, however, claim its employment when mixed with an exciting-fluid for the electrodes and applied thereto before being introduced into the cell, so as by chemical action to transform the zinc oxide at once into a chemically-different salt—as, for instance, dilute sulphuric acid—which transforms it into zinc sulphate.

"The inner resistance of the elements will not be raised by the addition of the oxide of zinc, as the latter is a better conductor of electricity than plaster and other similar bodies, which produce only a partial and varying porosity."

INVENTOR PROFILE

Alessandro Volta (1745–1827)

Alessandro Volta did not talk until the age of four. His family believed that he had an intellectual disability.

Alessandro Volta was an Italian physicist known for his pioneering work in the field of electrochemistry, which led him and others to astounding discoveries in electricity. He was born into a noble family in Como, Italy, in 1745. In 1774, he was appointed to his first academic position, as professor of physics at Como High School. The very next year he came up with the electrophorus, his first significant invention. The electrophorus was made up of two plates: One was metal and covered with ebonite, and the other plate featured an insulated handle. The device accumulated the charges created by static electricity, a process that is used as the basis of electrical condensers even today. Volta's most significant contribution to the field, however, was his invention of the first crude battery in 1800.

Volta made his discovery in the most unlikely manner: He learned of the work of a scientist friend, Luigi Galvani, who, while dissecting a frog, had noticed that the dead critter's leg would twitch when touched with two different metals. Volta was intrigued by Galvani's finding, and set out to determine whether the charge was coming from the animal's tissue or from the two metal pieces used in the dissection.

Beginning in 1794, Volta began a series of experiments to try to duplicate Galvani's work. He quickly determined that the living tissue wasn't necessary for the generation of electricity; instead, he discovered that the frog's leg was merely a conductor for a charge being generated by the two metal pieces themselves. They were oppositely charged metals, so they created an electrical impulse when they were connected, in this case by the frog's leg. Volta spent the next several years trying to refine this process, and in 1800, he developed the first electric battery, which came to be known as a Voltaic pile.

Volta's battery was a far cry from the small metal batteries we are accustomed to today. The Voltaic pile consisted of round plates of copper and zinc stacked up and separated by disks of cardboard moistened with salt water. A copper wire was attached to the top and bottom of this "pile," and when the wire ends were touched, electricity flowed through the stacked disks. The units of electricity that were created were dubbed "volts" after the inventor, a term that is still in use today.

Thanks to his groundbreaking invention, Volta received considerable acclaim in his lifetime. He received the Copley Medal of the Royal Society, and when the principles of the Voltaic pile became more widely known, he received many other medals and awards, including the French Legion of Honor. In 1801, he was made a count by Napoleon.

Bottle Cap

Patent Name: Bottle-Sealing Device
Patent Number: 468,226
Patent Date: February 2, 1892
Inventor: William Painter, of Baltimore, Maryland

What It Does

Provides an effective seal for bottled, carbonated beverages.

Background

Carbonated beverages were in popular demand by the late nineteenth century, but there were problems supplying a quality product in a bottle. Stoppers and other corking and capping implements used at the time did not reliably seal the bottles, and some had a tendency to either taint the flavor or flatten the carbonation. William Painter put a "stop" to these problems when he came up with his unique crown cap design in the early 1890s.

Painter's design featured not only the familiar crown shape of the cap that locked onto the lip of the bottle's neck; it also included compressible disks on the underside of the cap to provide a leak-free seal. But because his invention necessitated a change in the neck design of the bottle, it was not instantly accepted. Painter finally convinced beverage bottlers of its ingenuity, and his cap went on to set the standard for bottle caps, which have changed little since. A crafty inventor, Painter received patents on more than eighty inventions, but his crown cap remained his crowning achievement.

Like stamp collecting, crown bottle cap collecting is a real passion for many people, and the Internet has brought them all a little closer. Bottle cap collectors throughout the world maintain hundreds of websites that showcase rare crown caps or offer trade opportunities to other collectors.

Crown caps have found a niche in the art world, and are commonly used in sculpture and design. See www.interestingideas.com/out/cap.htm.

How It Works

Before plastic became the material of choice, as late as the 1960s, crown cap makers were still using cork for the caps' inner sealing material.

A bottle lip (a) is rounded, as is the exterior packing surface (a′).

1. The recess below the lip (c′)—the locking shoulder—extends at a short, straight incline.
2. Disk (C) or (C′) is placed on top of a filled bottle. Cap (B) is then placed on top and heavy downward pressure is applied.
3. Its flared edges (d) are bent downward and inward to engage with the locking shoulder.
4. The flange design provides access to free space (e) for convenient opening with the aid of a bottle-opening device (usually a prying tool that employs leverage against the top of the cap).

W. PAINTER.
BOTTLE SEALING DEVICE.
No. 468,226.   Patented Feb. 2, 1892.

In the Inventor's Words

Two years after receiving his bottle-cap patent, Painter was issued a patent for a "bottle cap lifter," just what was needed to pop the top.

"My present invention pertains to the sealing of bottles by the use of compressible packing—disks and metallic caps, which have flanges bent into reliable locking engagement with [ringed] locking-shoulders on the heads of bottles, while the packing disk is in each case under heavy compression and in enveloping contact with the lip of the bottle."

An American Arithmometer calculator from 1891

Calculator

Patent Name: Calculating-Machine
Patent Number: 388,116
Patent Date: August 21, 1888
Inventor: William Seward Burroughs, of St. Louis, Missouri

What It Does

A high-speed device for solving simple arithmetic problems.

Background

The inventor's grandson was the experimental and groundbreaking writer William Burroughs, whose many books include *Naked Lunch*.

This invention bears almost no resemblance to the common electronic or battery-powered calculators we use today. In size, it more closely resembles a typewriter. In function, it is really an adding machine that also multiplies by making repeated additions. But this patent represents the first successful printing calculator. Two years before the issuance of this patent, Burroughs and three associates established the American Arithmometer Company. They sold their first machine for a whopping $475, an almost unimaginable sum for such a device in the nineteenth century. But there was a pressing demand for what was then an extremely unusual instrument.

Previous versions of these machines were expensive, inefficient, and easily damaged. A common feature at the time was a handle that was pulled to register the numbers; varying pressures on these handles contributed to the breakage of many early adding machines. Burroughs's invention included a kind of hydraulic regulator on the handle so that no matter how sharply or gently the handle was pulled, pressure exerted on the mechanism was the same. Burroughs's calculator also printed out the calculation on a strip of paper which could be used as a record or a receipt for accounting purposes. This innovation had a major impact on the way office and business transactions were conducted.

How It Works

1. A keyboard consisting of nine columns of numbers in ascending order from 1 through 9 is arranged in nine rows.
2. The keys deploy revolving disks, which are fixed on a shaft, that contain on their peripheries figures 0 through 9.
3. An operator strikes the keys, and a series of mechanized operations tallies a cumulative figure.

In the Inventor's Words

William Burroughs was posthumously inducted into the Inventors Hall of Fame in 1987.

"My invention relates to that class of apparatus used for mechanically assisting arithmetical calculations; and my invention consists in the combination, with one or more registers, of a series of independent keys and intervening connections constructed, arranged, and operating, as fully specified hereinafter, so as to indicate upon the register the sum of any series of numbers by the proper manipulation of the keys, and also so as to print or permanently record the final result."

Cellophane

Patent Names: Process for the Continuous Manufacture of Cellulose Films; Apparatus for the Continuous Manufacture of Cellulose Films
Patent Numbers: 981,368; 991,267
Patent Dates: January 10, 1911; May 2, 1911
Inventor: Edwin Brandenberger, of Vosges, France

Q: "In two thousand years, what's the greatest invention you've seen?"
A: "In two thousand years? Saran Wrap."
—Carl Reiner and Mel Brooks, "The 2,000-Year-Old Man"

What It Does

Offers a method of producing a thin, strong, flexible, continuous sheet of transparent cellulose film that may be applied to a variety of practical uses.

Background

"You're the top.
You're Mahatma Gandhi,
You're the top!
You're Napoleon brandy.
You're the purple light of a summer night in Spain.
You're the National Gallery. You're Garbo's salary,
You're cellophane!"
—Cole Porter, "You're the Top"

When Mel Brooks, in his legendary comedy routine "The 2,000-Year-Old Man," jokingly paid homage to the impact of Saran Wrap, he wasn't far off the mark. But before there was Saran, there was cellophane, which opened the door to an array of innovations in plastics. At its inception, it was meant only to be waterproof.

One story has it that, after he observed a glass of wine spilled onto a restaurant tablecloth, Swiss chemist Edwin Brandenberger decided to work on developing a stainproof fabric. Employed at the time by a French textile company, Brandenberger coated cloth with a viscose film. But while no one was interested in his product, he realized he was onto something and continued experimenting with the liquid viscose, a cellulose solution derived from plants.

On July 23, 1909, he applied for two patents: one for the process of making the diaphanous film, and another for the apparatus used to make it. In 1911, he was awarded both of these patents.

Brandenberger combined the words "cellulose" and "diaphanous" to coin the term "cellophane." The first version of the product had a greenish hue and was too expensive to be used for anything but decorative packaging for upscale merchandise. Brandenberger later assigned his patents to an organization named La Cellophane Société Anonyme. This group sold patent rights to the DuPont Cellophane Company. Despite some notorious legal battles, DuPont secured a comfortable hold on the stuff, and DuPont scientists went on to perfect and market the product in a variety of forms. Dow Chemical Company soon jumped on the bandwagon, and it was this company that created Saran Wrap.

How It Works

1. A cellulose water solution, cellulose xanthate, is distributed through a hopper into a solution of sulfate and ammonia. Upon contact, the first solution coagulates.

2. The coagulated film is immediately treated in a water solution of sodium chloride or sea salt to remove impurities.

3. The film is immediately directed to a third bath consisting of a mineral acid diluted with water. In its final insoluble state, the film is washed in cold water and hot water.

4. The apparatus in which the whole process can be orchestrated consists of a series of vats and rollers, as depicted in the patent illustration.

991,267.

E. BRANDENBERGER.
APPARATUS FOR THE CONTINUOUS MANUFACTURE OF CELLULOSE FILMS.
APPLICATION FILED JULY 23, 1909.

Patented May 2, 1911.
2 SHEETS—SHEET 1.

WITNESSES

Fig.1

INVENTOR

BY

ATTORNEYS

In the Inventor's Words

"This invention, relates to a process by means of which it is possible to obtain in a continuous manner films of an indefinite length, starting direct from a water solution of cellulose, more particularly from a solution of cellulose xanthate."

Chewing Gum

Patent Name: Improved Chewing Gum
Patent Number: 98,304
Patent Date: December 28, 1869
Inventor: William F. Semple, of Mount Vernon, Ohio

What It Does

Dispenses a sugar-based "snack" while providing an active chewing experience as its mass remains in the mouth undigested.

Background

The chewing of gum was hardly an innovation in the nineteenth century. Ancient Greeks chewed *mastiche*, the resin of the mastic tree; Aztecs and Mayan Indians chewed chicle.

Thanks to the popularity of chewing gum in the late 1800s, the vending machine came into existence to dispense the product. Some of the first vending machines appeared in the New York subway system and offered only gum.

As a dentist, William Semple had the best intentions when he patented his invention: to encourage people to keep their jaws well exercised and their teeth clean. Recommended ingredients for his rubber-based "chewing gum" included chalk, charcoal, and powdered licorice root—ingredients that he believed would provide scouring properties. While Semple never aggressively marketed his invention, he did help lay the groundwork for others with names like Wrigley.

Meanwhile, the commander Antonio López de Santa Anna, who led the bloody charge against Fort Alamo before his army retreated, had settled on Staten Island in New York. He brought from his native land a large chunk of dried chicle, the gummy, tasteless resin of the sapodilla tree, which he enjoyed chewing. He introduced inventor Thomas Adams to his chicle; Adams imported a large amount with the intention of making an inexpensive synthetic rubber. His experiments failed, so he decided on improving the stuff as a chewing gum.

Less than two years after Semple received his patent, in 1871, Adams began selling his flavorless chicle gumballs in a New Jersey drugstore. Adams was also the first to sell gum that could be purchased from a vending machine. Soon flavors were added and the history of gum began to take on new dimensions. The good dentist Semple could not have known that what he started would lead to the subsequent generations of cavity-inducing, sugar-laced gum that have become a staple to youngsters worldwide.

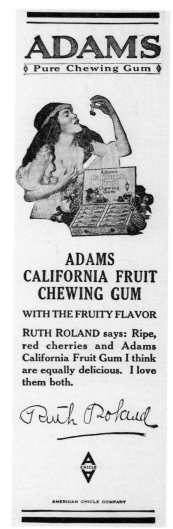

A 1919 advertisement for Adams California Fruit Gum, which was considered an exotic flavor.

How It Works

Semple's gum recipe is simple:

1. Dissolve the rubber in naphtha and alcohol until it reaches a jellylike consistency.
2. Mix in prepared chalk, powdered licorice root, or other suitable materials.
3. Enjoy, if possible.

In the Inventor's Words

"The nature of my invention consists in compounding with rubber, in any proportions, other suitable substances, so as to form not only an agreeable chewing-gum, but also, that from the scouring-properties of the same, it will subserve the purpose of a dentifrice.

"It is well known that rubber itself is too hard to be used as a chewing-gum, but in combination with non-adhesive earths may be capable of kneading into any shape under the teeth."

By the beginning of the Civil War, the Colt revolver was one of the world's most popular small arms.

Colt Revolver

Patent Name: Improvement in Fire-Arms
Patent Number: X9430
Patent Date: February 25, 1836
Inventor: Samuel Colt, of Hartford, Connecticut

"God didn't make all men equal. Sam Colt did."
—Popular expression in the Wild West

What It Does

This easily loaded weapon contains multiple shots, which can be discharged in rapid succession.

Background

Using electronic ballistics technology, Australian Mike O'Dwyer invented a weapon that can fire up to a million rounds per minute. A test firing of thirty-six barrels lashed together turned fifteen wooden doors into tiny shreds in less than a quarter of a second.

Samuel Colt's invention transformed a firearm from a single-shot device to a multiple-shot device. He developed a revolving chamber; the application of caps at the end of the cylinder; a partition between the caps; a shield over the caps to protect them from moisture and smoke; and the novel principle of a connecting-rod between the hammer and the trigger. The same year he received his first U.S. patent, he established a factory in New Jersey to make the guns. But it would be six years before Colt's gun earned more than just modest attention. In 1842, his Patent Arms Manufacturing Company closed its doors.

Colt turned his attention elsewhere and invented a device that could set off explosives by remote control. But he soon returned to guns. In 1847, the doors of his business reopened when the U.S. Army entered into a contract with him to produce one thousand of his revolving-cartridge firearms for use in the Mexican War. His guns soon became the most popular during the Civil War. Colt continued honing his products and eventually produced a metallic-cartridge revolver that substantially reduced a gun's tendency to misfire. This model was revered for its accuracy. It was called the Colt .45-caliber Peacemaker, and quickly became the gun of choice in the American West.

How It Works

1. A hammer acts upon a fulcrum, and is drawn back by a projection.

2. A pin projection from the hammer locks the cylinder into place when it is aligned with its respective chambers.

3. Pulling the trigger from the catch of the hammer pushes a spring mechanism forward.

4. The spring activates delivery of the percussion cap into a tube, where it explodes and discharges the load.

In the Inventor's Words

Samuel Colt was a true entrepreneur who understood the finer points of financing. Under the alias "Doctor Coult," he procured enough nitrous oxide to tour the country earning money demonstrating the effects of laughing gas.

"Among the many advantages in the use of these guns, independent of the number of charges they contain, are, first, the facility in loading them; second, the outward security against dampness; third, security of the lock against the smoke of the powder; fourth, the use of the partitions between the caps, which prevent fire communicating from the exploding cap to the adjoining ones; fifth, by the hammer's striking the cap at the end of the cylinder no jar is occasioned, deviating from the line of sight; sixth, the weight and location of the cylinder, which give steadiness to the hand; seventh, the great rapidity in the succession of discharges, which is effected merely by drawing back the hammer and pulling the trigger."

Drinking Straw

Patent Name: Artificial Straw
Patent Number: 375,962
Patent Date: January 3, 1888
Inventor: Marvin C. Stone, of Washington, District of Columbia

What It Does

Provides a means of ingesting beverages or medications through an elongated tube by the use of a sucking motion, rather than engaging directly with the container holding the liquid.

Background

Until the late 1880s, such natural grasses as rye or hollow reeds were used as drinking straws and as a means to administer medicine. Then straws of paper were patented for the purpose in England. The forerunner of today's drinking straw was invented by an American manufacturer of paper cigarette holders, Marvin Stone. In the days before synthetic plastic, it struck Stone that a better straw could be made by coating manila paper with paraffin and fashioning it into a tube. Eventually, a machine was devised to coil and wind the paper into tubes.

The spiral-winding technique that his company used to form the tubes may have been a more patent-worthy innovation than the straw itself. While plastic soon came to replace paper as the preferred material for drinking straws, Stone's spiral-winding method is now commonly used to make electronic components and countless other products in a variety of industries.

How It Works

1. Paper (A) is cut and formed with parallel sides and beveled ends (a, b).

2. A cylindrical spindle containing a slit (d) to receive an edge of the paper winds the paper in a spiral motion to form the tubing.

3. The tube is secured by adhesive (e) pre-applied to the opposite edge of the paper.

4. The tube is then dipped into a vat of molten paraffin to seal it and make it durable against the liquid into which it will be inserted.

In the Inventor's Words "The aim of my invention is to provide a cheap, durable, and unobjectionable substitute for the natural straws commonly used for the administration of medicines, beverages &c.; and to this end it consists, essentially in a straw formed by winding a paper strip into tubular form and securing the final or outer edge by an adhesive material, the whole being coated with paraffin or other water-proof material."

Levi Strauss Blue Jeans

Patent Name: Improvement in Fastening Pocket-Openings
Patent Number: 139,121
Patent Date: May 20, 1873
Inventor: Jacob W. Davis, of Reno, Nevada, assignor to himself and Levi Strauss & Company, of San Francisco, California

What It Does

Durable, comfortable pants, ideal for physical labor, include extremely sturdy seams, including those around their pockets, which are enforced with rivets.

Background

Before denim, the textile used in making sturdy pants typically came from Genoa, Italy. Weavers in France called the material *genes*, which subsequently translated into "jeans" in English. Similarly, when the Italian-made textile was replaced by a softer fabric manufactured in Nîmes, France, it became known as the fabric *de Nîmes*, or in American-speak, "denim."

Levi Strauss & Company considers the official birthday of blue jeans as we know them today to be May 20, 1873. In reality, denim jeans were around long before then, regularly used to make durable work-wear. Levi Strauss, a native of Germany, had already established himself as a successful entrepreneur. His company manufactured denim blue jeans in San Francisco during the mining boom years of the western United States.

A Reno-based tailor from Latvia named Jacob Davis regularly ordered bolts of cloth from Levi Strauss & Company. As the story goes, one of Davis's clients returned regularly with complaints about shoddy seams, which were evidently prone to ripping. The crafty tailor solved the problem very simply with metal rivets. The small metal reinforcements could be fastened into the denim at the most critical junctures, such as the pocket corners. Davis's idea fared well with his customers, but he couldn't afford to shell out the sixty-nine bucks required to patent his idea. In need of financial backing, he partnered with Levi Strauss. The rivet concept is still used by the company, but extra-durable stitching is often substituted, partly as a concession to fashion. Far from being strictly for miners and other laborers, blue jeans have become the most popular form of casual dress in the world.

How It Works

A metal rivet is studded through a hole at the end of a seam, binding two pieces of cloth together securely. It is then "headed down," or compressed on both sides, firmly securing the seam.

In 1936, Levi Strauss & Company introduced the brand-recognition concept by sewing its little red signature flag along the back pocket of its jeans. It was the first label sewn on the outside of a garment.

Dubbed the "XX," the world's oldest pair of Levi Strauss & Co. jeans is believed to have been made in 1879 and is stored in a fire-proof safe at the company's archives in San Francisco. Their estimated worth is more than $150,000.

J. W. DAVIS.
Fastening Pocket-Openings.
No. 139,121.   Patented May 20, 1873.

Fig. 1.

Witnesses   Inventor

In the Inventor's Words "My invention relates to a fastening for pocket-openings, whereby the sewed seams are prevented from ripping or starting from frequent pressure or strain thereon."

Optical Fiber

Patent Names: Fused Silica Optical Waveguide; Method of Producing Optical Waveguide Fibers
Patent Numbers: 3,569,915; 3,711,262
Patent Dates: May 2, 1972; January 16, 1973
Inventors: Donald B. Keck, Peter C. Schulz, and Robert D. Maurer, assigned by Corning Glass Works of Corning, New York

What It Does By means of successive internal reflections, this fiber transmits information through light beams.

Background

One of the original uses for optical fiber was a medical one. The fiber was used in endoscopes, tubes that can retrieve views from inside the human body.

Optical fiber was a revolutionary invention that picked up where Alexander Bell had left off in 1880. Way ahead of his time, Bell had invented what he called a "photophone," which transmitted sound on a beam of light. The invention was a direct precursor to fiber optics used in telecommunications today. Bell, however, was unable to resolve problems associated with light source interference and so could not refine his creation. Optical fiber would prove to be the perfect mechanism for what Bell had first conceived, and for much more. A flexible, transparent fiber, usually made of glass or plastic, transmits information by successive internal reflections.

On May 11, 1970, three researchers with the Corning Glass Works filed applications for two patents: one for the optical fiber itself; the other for a method to produce it. Earlier that year, they had produced an optical fiber that possessed more than 60,000 times the data-carrying capacity of copper wire.

The three researchers were not the only ones aware of the massive potential of fiber optics. Engineers had been trying to perfect it for some time, but were unable to resolve a fundamental problem: sustaining a significant amount of light as it traveled through the fiber. The Corning trio came up with a unique waveguide innovation that did the trick. This would prove to be a monumental achievement, which was first applied to long-distance telephone communications and ultimately changed multimedia telecommunications across the globe.

Light travels through optical fiber more efficiently than through the atmosphere.

How It Works

1. Production of fiber optics begins with a rod of fused silica doped with tin oxide, titanium oxide, or several other dopant materials that increase the index of light refractivity.
2. The rod is inserted into a tube of fused silica. The temperature of the rod-and-tube combination is then raised until it has enough viscosity to be drawn into a wire or fiber.
3. A transmitter produces and encodes light signals. The signals travel through the core (30) of the waveguide (10) in figures 1 and 2.

4. They are refracted by the mirrorlike cladding (20) made of pure fused silica or slightly-doped fused silica. An optical receiver decodes the data traveling in the light signals.

Jan. 16,   3,711,262

METHOD OF PRODUCING OPTICAL WAVEGUIDE FIBERS

Filed May 11, 1970

Fig. 1

Fig. 2

Fig. 3

Fig. 4

INVENTORS.
Donald B. Keck
Peter C. Schultz
·BY
Walter S. Zebrowski
ATTORNEY

In the Inventors' Words   "Fused silica has excellent light transmission qualities in that absorption of light energy and intrinsic scattering of light by the material is exceptionally low. Scattering of light that does occur within fused silica is primarily caused by the presence of impurities rather than the intrinsic nature of the material itself. Furthermore, fused silica is such a hard material that an optical waveguide once formed possesses the quality of being highly resistant to damage from high temperatures, corrosive atmosphere and other severe environments."

PEZ Dispenser

Patent Name: Pocket Article Dispensing Container
Patent Number: 2,620,061
Patent Date: December 2,1949
Inventor: Oskar Uxa, of Vienna, Austria, assignor to Eduard Haas, of Muhlbach-Attersee, Austria

What It Does   Offers a novel means to dispense small rectangular candies from a spring-loaded plastic container, appealing especially to young children for its toylike quality, and to collectors for the variety of characters depicted.

Background   In 1927, Austrian Eduard Haas III enlisted the help of a chemist to devise a cold-pressing process to make peppermints inexpensively. They designed the mints in a rectangular shape that allowed for more efficient machine-wrapping. Haas began marketing his peppermint tablets as an antismoking aid. The name PEZ comes from an abbreviation for the German word *Pfefferminz*, or peppermints. Perhaps to further its appeal as an antismoking aid, dispensers mimicking the action of lighting a cigarette were patented two decades later.

   Haas's company thrived, and when it was time to market his merchandise in North America in 1950, ad campaigns targeted kids and soon the dispensers had figure heads on them. Among the first characters to appear on PEZ dispensers were Santa Claus and Popeye. The pretense that PEZ was anything more than candy in a novelty dispenser stayed in Europe. It was the novelty that Americans couldn't resist. Today, vintage PEZ dispensers depicting characters that range from Barney Rubble to Barbie are considered hot collectibles.

Emoji Pez dispensers at the Pez Memorabilia Museum, which has every dispenser ever sold

How It Works   1. An oblong plastic container with an opening at the top and a moveable base houses a stack of tablets.

2. Surrounding this container is an outer casing with a spring at its base that is set to act against the moveable base of the inner container.
3. A pin-spring hinge at the top is set to open easily upon pressure of the thumb and to snap shut when released.
4. When the thumb-piece is engaged, a thrust member inside dispenses a single tablet through the opening. The removal of a tablet automatically triggers the spring to push the stack of tablets upward by the inner container's moveable base.

Dec. 2, 1952   O. UXA   2,620,061

POCKET ARTICLE DISPENSING CONTAINER

Filed Oct. 14, 1949

Fig.1   Fig.2

Fig.3

OSKAR UXA, INVENTOR.

BY

ATTORNEY

In the Inventor's Words   "The invention relates to pocket containers having a hinged lid and adapted for the delivery of articles, such for example as pastilles, tablets, sweetmeats, and cigarettes, which containers can be opened or closed with one hand, and present the merchandise to the user in a hygienic and unspoilt condition. The possibility of operation by one hand only is important not only for persons having only one hand but also persons who often have only one hand free (for example motor-vehicle drivers), or whose occupation causes their hands to become smeared with dirt."

John Bardeen,
William Shockley,
and Walter Brattain

Transistor

Patent Name: Semiconductor Amplifier
Patent Number: 2,502,488
Patent Date: April 4, 1950
Inventor: William Shockley, of Madison, New Jersey, assignor to Bell Telephone

What It Does   Controls and manipulates electrical signals to maximize energy efficiency and conserve space.

Background

Texas Instruments produced the first practical silicon transistors in 1954. They were used in the first fully transistorized radios to hit the market. TI also marketed the first handheld calculator.

In conjunction with another patent (2,524,035) awarded to William Shockley's two colleagues John Bardeen and Walter Brattain for a Three-Electrode Circuit Element Utilizing Semiconductive Materials, this patent contributed to the creation of a solid-state device that could amplify electrical current—the transistor. For this great contribution, the three inventors shared a 1956 Nobel Prize. The transistor revolutionized modern electronics—replacing the vacuum tube—and became a key component in microchip and computer technology. Until the transistor's appearance, vacuum tubes were the only way to amplify electronic signals or serve as switching devices. Smaller, lighter, more durable, and more reliable than vacuum tubes, the transistor would replace the vacuum tube's function and signaled a new era in electronics technology in general, radio technology specifically, and computer technology ultimately.

But the transistor was not an overnight success. For many years, the transistor was simply not selling very well. Its price, higher than the vacuum tube, was apparently greater than its appeal. Eventually though, the trend toward miniaturization in electronics marginalized the vacuum tube as the transistor grew smaller and cheaper. This proved to be a turning point in history, when the science-fiction concept of computers became a very real feature of the workplace. Today a large portion of the world population would be unable to imagine life without computer technology. The transistor, in fact, is the technological father of the integrated circuit, which consists of microchips that link together millions of tiny transistors.

How It Works   The transistor functions dually as a tiny on/off switch and as an amplifier of electricity. Like a water faucet valve, it turns the current on and controls the rate at which it flows. A base, a collector, and an emitter make up the three basic parts of the transistor. The base acts like the control valve, opening and closing access to a larger supply of electricity provided by the collector and dispersing it through the emitter like a faucet tap.

In the illustration, a block of semiconductive material is depicted.
1. Separating the two types of electricity carriers in semiconductors (the p-type and the n-type) is a high-resistance electrical barrier (12).
2. Large area connections (13, 14) at both ends of the block are formed of solder or electroplated metal coating.
3. A conductive point contact (15) is adjacent to the barrier.
4. Current introduced by the contact causes electrons to gather and form a more efficient channel of electricity.

The materials that make up the transistor, conductors, and insulators have been essential to its success. Conductors transfer electrical current; insulators do not. The material used for the base of today's transistors—pure silicon—is considered a semiconductor. The trio of inventors knew that the semiconductive properties of silicon, or germanium, could be exploited to create a more efficient and key component in electronics.

In the Inventor's Words "This invention relates to means for and methods of translating or controlling electrical signals and more particularly to circuit elements utilizing semiconductors and to systems including such elements.

"One general object of this invention is to provide new and improved means and methods of translating and controlling, for example, amplifying, generating, modulating, etc., electric signals.

"Another general object of this invention is to enable the efficient, expeditious and economic translation or control of electrical energy."

CHRONOLOGY OF THE COMPUTER AGE

Computer technology has progressed exponentially in the last thirty years or so, with new machines and processors coming out so fast that some machines are rendered obsolete just a few years after being introduced to the market. However, the early development of computers did not progress at such lightning speed. The first forerunners of modern-day machines were clunky, slow behemoths, and until only recently, it seemed that the giant calculating machines would be used only by trained professionals in giant laboratories.

At nineteen, Blaise Pascal invented the first calculator.

◀ **1642:** The first crude calculator—a set of wheels linked by gears—was invented by Blaise Pascal, a French mathematician, philosopher, and scientist.

Early 1670s: Gottfried Wilhelm Leibniz, a German mathematician, improved Pascal's calculator, adding components that allowed the machine to do multiplication and division. Leibniz also developed the binary system, which provided a shortcut to performing decimal-based operations.

Mid-1800s: English logician and mathematician George Boole refined the binary system, which represents data with just two numbers, 0 and 1. To this day, incredibly complex calculations are performed using software based on what is now referred to as "Boolean logic."

1801: Joseph-Marie Jacquard developed a punch-card system to automate looms. A card with a pattern created by punched-out holes was used to direct needles in the weaving process. The holes (and absence of holes) replicated the two digits—the 1s and 0s—of the binary system.

Charles Babbage

1830s: English mathematician Charles Babbage used punch-card technology to develop a mechanical computer that stored sets of punched cards for later reference in complex operations.

1890: Herman Hollerith, an American inventor and businessman, created a machine that used the punch-card system to tabulate the results of the 1890 census. His is considered the first successful computer.

1914: International Business Machines Corporation (IBM) was founded. The company is an outgrowth of Hollerith's Tabulating Machine Company.

1944: Harvard University professor Howard Aiken built the Mark 1, a digital computer controlled by electromechanical switching devices.

1945: J. Presper Eckert, Jr., and John William Mauchly, both engineers at the University of Pennsylvania, developed ENIAC, the Electronic Numerical Integrator and Computer. The giant computer took up over 15,000 square feet and was operated by 18,000 vacuum tubes.

William Shockley

1947: Scientists at the Bell Telephone Laboratories, led by William Shockley, invented the transistor. The transistor was smaller, lighter, more durable, and more reliable than vacuum tubes, and its creation signaled a new era in computer technology. The transistor's main function is to rapidly turn on and off the flow of electricity, and to amplify the flow when needed.

1951: Eckert and Mauchly joined forces with John von Neumann, an early pioneer in the field of computer logics, to develop EDVAC, the Electronic Discrete Variable Automatic Computer, an influential precursor to later computers. Later that year, Eckert and Mauchly also completed the UNIVAC, the Universal Automatic Computer, which became the first commercially viable computer.

1958: American engineer Seymour Cray designed the first computer powered entirely by transistors.

1959: IBM unveiled its first transistorized computer.

1959: The earliest integrated circuits were introduced, otherwise known as computer chips (or silicon chips). The miniaturization of computer technology began.

Late 1960s: Businesses began to rely more heavily on computers and link them into networks.

Late 1960s: The U.S. government's Department of Defense developed a way to link all of its computers together as a preventive measure in case of attack or natural disaster. The network was called ARPAnet. Eventually, businesses, universities, and other institutions linked their own systems into this nationwide network, creating what is known today as the Internet.

Early 1970s: Computers became smaller as developments in chip technology made it possible for more and more information to fit into smaller and smaller spaces.

The Altair 8800

January 1975: The Altair 8800 computer—a build-it-yourself kit based on the Intel 8080 processor—was sold via mail order by Micro Instrumentation and Telemetry Systems (MITS), a company in Albuquerque, New Mexico. *Popular Electronics* magazine made it a cover story. The name comes from the television show *Star Trek* and was suggested by the daughter of MITS's owner, Ed Roberts.

An older IBM computer with the MS DOS operating system

1977: Steven P. Jobs and Stephen G. Wozniak were intrigued by the Altair, but couldn't afford one, so they built their own. They went on to develop the Apple II personal computer. The relatively inexpensive machine was easy to use and practical for regular people who didn't have specialized computer training. Small business, families, students, and schools began to purchase computers.

1981: IBM introduced its version of the personal computer. It soon surpassed the Apple II in popularity.

1984: Apple bit back with the Macintosh, a simple-to-use desktop computer embraced by the public for its easy-to-use graphics and design programs.

1991: Englishman Tim Berners-Lee developed hypertext markup language (HTML), then used the new code to launch the World Wide Web. The Internet introduced graphics, video, sound, and other features to the Internet for the first time. Berners-Lee also devised the universal address system used on the Web, otherwise known as the Uniform Resource Locator, or URL, which gives each website its unique address.

Tim Berners-Lee invented the World Wide Web and ensured it would be free and available to all.

1993: Marc Andreessen, a student at the University of Illinois, began development of Mosaic, the first Web browser, designed to easily display HTML documents. Mosaic formed the basis of the popular Netscape browser Andreessen went on to develop in Mountain View, California.

Late 1990s: The World Wide Web became an immensely popular common medium used by people around the world. E-mail, Web pages, online retail, and chat rooms all became part of the popular vernacular, and in 1998 Google was founded.

2003: AMD's Athlon 64, the first 64-bit processor, was introduced to the consumer market.

2007 and beyond: Apple's Steve Jobs introduced the iPhone in 2007, which introduced many computer functions to the smartphone, and 2010 unveiled the iPad, jumpstarting tablet computers. With the development of 3G technologies, people can easily access the Internet no matter where they go.

Velcro

Patent Name: Velvet Type Fabric and Method of Producing Same
Patent Number: 2,717,437
Patent Date: September 13, 1955
Inventor: George de Mestral, of Prangins, Vaud, Switzerland

What It Does

This self-fastening material is made of tiny hooks on one side and loops on the other, which can be separated and reattached many times.

Background

Velcro is not used strictly for practical purposes. It is, for example, an important component in such human-performed stunts as the favorite run-jump-and-stick-trick. This involves a suit of Velcro and a Velcro-covered wall.

In 1948, the Swiss inventor George de Mestral returned from a walk in the woods with his dog to find himself and the dog covered in cockleburs. Intrigued by what made them stick, he looked at the burrs under a microscope and saw the natural loop-grabbing hooks protruding from the core. He decided to invent a fabric based on the adherent characteristics of the burr. He teamed up with a French weaver and, several years later, perfected what he called Velcro, a combination of the words "velour" and "crochet."

One of the most important little inventions in textile history, Velcro has found countless practical uses. In addition to replacing zippers and buttons on

Today, "Velcro" is both a trade-marked product and the name of the company that makes it.

clothing, shoes, and wallets, Velcro has endless more pragmatic purposes. Its appeal lies in its ability to stick together instantly and be torn open quickly—accompanied by a satisfying sound. It is even used to keep equipment in place aboard space shuttles.

How It Works

1. A foundation material is constituted by a weft strand (1) and a warp strand (2).
2. Warp threads (2, 3) are woven to form a raised pile (9, 10).
3. Threads (9) are bent down at the ends, forming a hook (4).
4. When the hook side meets the pile side of the Velcro, the hooks stick into the pile, forming a tight and tidy seal.

In the Inventor's Words

"My invention has for its object a velvet fabric including a foundation structure constituted by a weft and a warp incorporating threads that are cut at a predetermined length so as to form a raised pile. My novel fabric distinguishes from the other similar fabrics by the fact that the raised pile is made of artificial material, while at least part of the threads in said pile is provided near its end with material-engaging means, as required for adhering to a similar fabric or for scouring purposes."

Zipper

Patent Name: Clasp Locker or Unlocker for Shoes
Patent Number: 504,038
Patent Date: August 29, 1893
Inventor: Whitcomb L. Judson, of Chicago, Illinois

What It Does

A strip made of metal or synthetic material features "teeth" that interlock when drawn together, providing a simple and useful fastener for articles of clothing, handbags, or other fabric items.

Background

Like many great inventions, the zipper came about as a response to a personal problem.

Whitcomb Judson was inspired to invent the mechanism as a result of his large physique. Simply put, Judson had difficulty tying his own shoes.

At the 1893 Chicago World's Fair, Judson showcased his clasp locker mechanism on a shoe. Apparently, not many people were impressed. It was considered clunky and displeasing to the eye. The interlocking clasps were far apart, and the ring to operate the mechanism was rather large. As a replacement for the normal buckle or string system, the clasp locker was just too conspicuous. Nevertheless, the mechanism sewn into the unsightly footwear displayed some of the defining principles of the modern zipper.

How It Works

1. Hooked clasps with interlocking parts engage only when at an angle to the line of strain.

2. Overlapping or underreaching projections at the forward ends of the clasps prevent disengagement of the hooks.

3. A moveable guide that includes cam-ways for the clasps allows for the engagement or disengagement of the clasps in a single motion.

In 1917, electrical engineer Gideon Sundback received a patent for a "separable fastener," which was a considerable improvement upon Judson's. It included an increased number of interlocking parts per inch, set in two rows of teeth.

In 1851, Elias Howe received a patent for an earlier zipper-like clothing enclosure. Had he marketed his device, it might have secured his place in history as the creator of *two* noteworthy inventions: Howe had already invented the sewing machine.

A Japanese company called Yoshida Kogyo Kabushikikaisha became the largest supplier of zippers. Hence, the "YKK" engraved on zippers of all kinds. The world's largest zipper factory is in Macon, Georgia, and produces five million zippers a day.

In the Inventor's Words "The invention was especially designed for use as a shoe fastener; but is capable of general application wherever clasps consisting of interlocking parts may be applied, as for example, to mail-bags, belts, and the closing of seams uniting flexible bodies."

Zippo Lighter

Patent Name: Pocket Lighter
Patent Number: 2,032,695
Patent Date: March 3, 1936
Inventors: George Gimera and George G. Blaisdell, of Bradford, Pennsylvania, assignors to Zippo Manufacturing Company

What It Does

This portable, pocket-sized, reusable device provides flame, particularly for igniting smoking materials such as cigarettes or cigars.

Background

Since the company's beginnings, Zippo has made more than 500 million lighters.

War correspondent Ernie Pyle wrote about the high demand for the Zippo in World War II, calling it "the most coveted thing in the Army."

The first Zippo lighter was introduced in 1932 and cost the consumer $1.95. This was a hefty price to pay in the days of the Depression. But George Blaisdell, later known as Mr. Zippo, was determined to manufacture and market a quality product. One of his brilliant marketing tactics was to offer a lifetime warranty. In a time before the majority considered smoking a bad habit, a lifetime warranty on a sharp-looking, windproof lighter was an attractive feature.

Today, the Zippo lighter says two things of its owner: You take your smoking seriously, and you take your style seriously. When a stranger at the train depot asks for a light, nothing is cooler than responding with the nonchalant flip of a Zippo lid and an instantaneous snap of flame. With its chic and compact design, the refillable lighter also becomes a personal keepsake.

How It Works

1. A casing houses two disposed hollow members (1, 2) that fit snugly against one another.
2. A top plate (3) encloses the inner housing.
3. At the top center of the plate, an elliptical windscreen with draft openings (6) emerges.
4. On one side, a toothed wheel (9) is supported to engage an elongated flint (11).

5. A small hinge (19) at the side of the top plate opposite the toothed wheel attaches to the cover (18) of the outer casing.

6. When the lid is opened, a downward flick of the thumb on the wheel creates a spark against the flint. The spark is ignited into flame by a wick that extends from within the lighter's interior and the lighter fluid tank.

7. Light up and take a drag. Don't inhale the first drag: You don't need that butane.

March 3, 1936.   G. GIMERA ET AL   2,032,695
POCKET LIGHTER
Filed May 17, 1934

In the Inventors' Words   "Pocket lighters having covers hinged on their upper ends must have means for holding them closed if they are to be satisfactory, and it is also desirable that they have means for preventing the covers from closing prematurely and extinguishing the flame after they are opened. In lighters known heretofore these means take the form of exposed latches and interior springs and levers which take up space in the lighters and reduce their storage capacity for inflammable lighter fluid. Furthermore, the exposed latches are apt to catch in and wear the clothing, to be accidentally actuated in the pocket, and to accumulate dirt. Such lighters are complex and frequently get out of order, while they must be constructed in a cheap manner in order to be saleable at a low price."

Three

The More Humane Patent

Adhesive Bandage

Patent Number: 2,823,672
Patent Date: February 18, 1958
Inventors: Peter Schladermundt, of Bronxville, New York, and William H. Dennerlein, of Beechhurst, New York; assigned by Johnson & Johnson

What It Does

Band-Aid bandage strips are used around the world to treat minor burns and cuts. The sterile gauze pad protects the wound and absorbs blood while promoting healing.

Background

The Band-Aid story actually goes back thirty-eight years before this patent was issued, to a young newlywed couple living in New Brunswick, New Jersey, in 1920. Earle Dickson worked for Johnson & Johnson, a company that specialized in the manufacture of surgical dressings. Josephine was a housewife prone to accidents in the kitchen. Dickson found himself regularly tending to her cuts and burns when he was struck with the idea of placing pieces of gauze onto strips of tape in anticipation of his wife's next mishap.

A 1952 Band-Aid ad from Johnson & Johnson

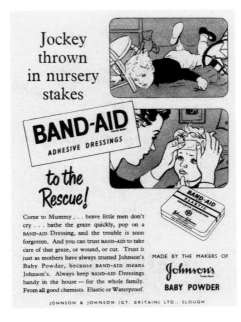

Dickson soon convinced his employers that the idea of these ready-made bandages might have commercial appeal; Johnson & Johnson began making the first adhesive bandages, cautiously, in 1921. The handmade strips were 3 inches wide and 18 inches long, and did not enjoy overnight success—but eventually, the idea stuck. In 1924, the company installed machines for mass-producing the new product, and the trade name Band-Aid was adopted. Dickson was made a vice president of the company, presumably allowing him to take his wife out to dinner a little more often.

In 1951, the first Band-Aid brand plastic strips were introduced. This patent is a close cousin to what most of us would recognize as the Band-Aid we use now. Specifically, this patent solves some of the previous problems of the product (e.g., the peel-away strips had too much contact with the adhesive around the sterile pad, causing potential for inadvertent contact between the pad and a person's nonsterilized finger).

How It Works

1. An adhesive bandage has a top-facing strip and a bottom-facing strip.
2. The bottom strip overlaps absorbent sterilized gauze.
3. A backing strip contains adhesive on its top.
4. The gauze pad divides the backing strip into two exposed areas on opposite endwise portions of the strip, which can be quickly peeled apart at the time of application.

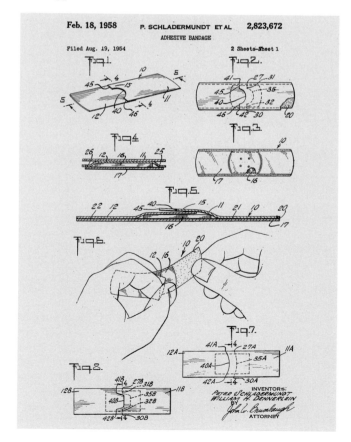

"Adhesive bandages, the term being used herein to designate not only the elongated strips, but also the round 'spots' and the relatively square 'patches,' are formed of a backing sheet which may be cloth or plastic coated on one side with an adhesive and having a dressing pad secured to the adhesive, leaving exposed areas of adhesive on both ends of the dressing. The endwise adhesive tabs generally constitute the major area of exposed adhesive. However, even in the elongated strips appreciable areas of exposed adhesive often appear at the sides of the dressing, although the latter are generally smaller, relatively speaking, than the areas of adhesive exposed at the ends of the dressing. The sidewise exposed adhesive areas are proportionately larger in the case of the round 'spot' and the square 'patch' dressing, in which case, they are usually as large as the endwise adhesive areas. Accordingly, the present invention is directed to the 'spot' and 'patch' dressings as well as the elongated strip adhesive bandages."

Air Bag

Patent Name: Air Bag Restraint System with Venting Means
Patent Number: 5,071,161
Patent Date: December 10, 1991
Inventors: Geoffrey L. Mahon, of Ridgewood, New Jersey, and Allen Breed, of Boonton Township, New Jersey

What It Does

Prevents injuries and saves lives during automobile accidents by inflating upon impact and creating a buffer between driver or passenger and the car's interior, windshield, etc.

Background

In 1968, Allen Breed invented an automotive sensor and safety system that employed the world's first electromechanical automotive air bag system. By 1973, GM was testing some of its first air bags. The automotive industry was already well aware of mounting consumer concerns over driver and passenger safety, but public demand was not so great as to demand that air bags become a regular feature in automobiles, and the industry was in no hurry to add more expense to its overhead.

When, eventually, the air bag did become a standard feature, it saved many lives. But injuries arising from the air bag itself were commonly reported as well. Tests and reports by accident victims revealed that, in many cases, the pressure of the air bag upon deployment was too high. The bag, meant to cushion the impact of a crash, was creating its own forceful impact. Using the resources of his automotive technology company, Breed modified his invention by adding a ventilation system to his air bag that would buffer the velocity of the impact while still protecting the passenger. Warnings were added against subjecting small children or the frail to the impact of the air bag.

How It Works
1. Crash sensors are located in several areas on the front of a vehicle.
2. The sensors connect to an air bag assembly through a wiring harness.
3. During a collision, the sensors send an electric spark to an inflator canister.
4. Within the space of a second from the time the sensors go off, nitrogen gas inflates the air bag at a rate of between 130 and 200 mph.
5. Vents in Breed's improved air bag allow slight deflation of the gas at a preselected rate to eliminate injuries from secondary collisions of a passenger with the bag.

"In an ideal situation, a constant restraint load is applied at the instant the impact begins and is such that the occupant utilizes all the available distance within the occupant compartment. Such a force would be the minimum required to dissipate the occupant's energy.

"In the worst case, no restraint force is applied until the vehicle is fully deformed and has come to rest, and the occupant has completely traversed the compartment. At this point the occupant will then impact the steering wheel, windshield and/or instrument panel. Depending upon the compliance of these components (which, in general, are not very compliant and also tend to exert localized loading), the occupant will sustain his energy change by way of a very high force exerted over the relatively short distance these components will yield.

"What can realistically be achieved lies somewhere in between these two extremes."

Air Conditioning

Patent Name: Apparatus for Treating Air
Patent Number: 808,897
Patent Date: January 2, 1906
Inventor: Willis H. Carrier, of Buffalo, New York

What It Does

Decreases the temperature of indoor air by drawing in air from outdoors, cooling it, and then circulating it throughout a room by means of fans and blowers.

Background

The company that Carrier started in 1915 is now an affiliate of United Technologies and conducts business in more than 170 countries.

Willis H. Carrier's air-conditioning epiphany came while he waited for a train one uncomfortably humid evening. A moment of clarity revealed to him the correlation between temperature, humidity, and dew point. He soon created the first air conditioner. Carrier's invention was first put to use in a Brooklyn printing plant, where fluctuations in temperatures warped the paper. As a result, a frustrated printer couldn't get the colors aligned.

After this success, the young engineer was awarded his own subsidiary company to make the device for businesses and warehouses, where large machinery required cooling or the products they made required a way to keep them from melting. Countless product lines suffered as a result of extreme and varying conditions in the air. The textile industry became a major supporter, since certain fabrics were prone to deterioration in uncontrolled temperatures. Air conditioning was especially useful in cotton mills, where a lack of moisture caused static electricity in the fibers, which were difficult to weave.

In 1915, Carrier and a small group of others established Carrier Engineering Company, which promised to deliver systems that could regulate temperature and humidity levels as desired by their industrial customers.

It was not until 1924 that Carrier's ideas were applied to cooling air for personal comfort rather than industrial needs. Many people felt their first blast of AC in department stores and movie theaters, which had installed them to keep their customers coming in the summertime.

The Depression slowed the success of the air conditioner, but not for long. The postwar housing sprawl that popularized a suburban lifestyle also boosted the demand for air conditioning. Today, for most people living in developed nations, it is hard to imagine life without the luxury of a regulated indoor atmosphere.

Willis Carrier

How It Works

1. A fan or propelling device (K) sends air current through the air trunk (M).

2. Attached to a supply pipe (F), a spraying device (H) is situated near the air trunk intake.

3. Sprayed air travels through a separator made of baffle plates to form a series of deflections to remove moisture and impurities in the treated air.

4. Flanges on the plates (b, c) keep the moisture trapped against the plates while the air continually makes contact with it in its sinuous path through the separator.

5. A basin (J) and filter (L) collect the moisture at the bottom. Inside the casing, pipe coils (B) provide a heating or cooling mechanism to control the air temperature.

No. 808,897.

PATENTED JAN. 2, 1906.

W. H. CARRIER.
APPARATUS FOR TREATING AIR.
APPLICATION FILED SEPT. 16, 1904.

Fig. 1.

Fig. 2.

Fig. 3.

Fig. 4.

Fig. 5.

Fig. 6.

Witnesses.
E. A. Volk.
R. W. Renner.

Inventor.
Willis H. Carrier
by Wilhelm Parker & Hard
Attorneys

In the Inventor's Words "This invention relates to apparatus for treating air previous to its use for ventilating and heating buildings or for other commercial purposes—such as drying, refrigerating, &c.—and more particularly to air-purifying apparatus of that kind in which a liquid or solution in a finely-divided condition or atomized spray is introduced into a current of air to be treated, which is then caused to pass through a separator consisting of baffle plates which intercept and separate from the air the particles of liquid, together with the solid impurities contained therein."

Men constructing
artificial legs c.1920

Artificial Leg

Patent Name: Artificial Leg
Patent Number: 37,282
Patent Date: January 6, 1863
Inventors: T. F. Engelbrecht, of New York, New York; R. Boeklin, of Brooklyn, New York; and W. Stahlen, of Brooklyn, New York

What It Does

Provides a viable, comfortable, and aesthetically acceptable substitute for a lost leg.

Background

In 1946, the development at the University of California, Berkeley, of a suction sock for above-knee prosthetics was a milestone in the attachment of lower limbs.

As represented in the model photo, this Civil War prosthetic may look like part of an early design for the flamboyant *Star Wars* robot C-3PO. But this improvement on artificial limbs was a significant ancestor to modern prosthetics, and it proved extremely useful during the Civil War, a bloody time in American history that claimed the lives of more than 600,000 and made amputees of some 100,000 more. Up until the time that artificial limbs of this sort were available, amputees had to rely on ill-fitting, often painful, and usually unsightly "peg legs."

How It Works

1. The thigh-piece (A) can be laced to fit closely onto the stump of the natural leg while simultaneously allowing ventilation to the healthy tissue.
2. The thigh-piece connects to the lower leg (B) and upper portion of the ankle joint (C), to which the foot (D) is connected.
3. The knee joint is replaced here by means of a joint pin (E); and a spring (G) is applied to straighten the knee joint, mimicking the movement of a biological limb.
4. A ball-(J)-and-socket ankle joint is employed with a spring to allow the foot independent movement while maintaining its proper position in relation to the leg

There are many complex and ingenious principles employed in this patent, both in its cost-effective production and in the product itself. For example, the design even accounted for subtle variations in the arch of the foot. The inventors had come up with an adjustable artificial limb that could be mass-produced at low cost while offering a strong and lightweight substitute for a missing limb. They proposed using sheet metal wrought into shape for the thigh and lower leg, connecting the edges by soldering. Such thinking advanced the field of modern prosthetic science.

In the Inventors' Words "This invention consists in certain provisions for the adjustment of the parts of an artificial limb in such manner as may be desirable to adapt it to the length of the natural limb and conformation of the foot of the intending wearer, by which means the necessity of making a limb to suit each particular case is to great extent obviated, and in consequence the cost is considerably reduced; also, in a certain novel construction of the ankle-joint, by which the movements permitted to the said joint are made to approximate nearer to the natural joint than it does in the artificial legs heretofore constructed, such construction being also applicable to the wrist and other joints of an artificial arm and hand; and it further consists in the construction of an artificial limb as are required to be rigid, of corrugated plate metal, whereby the limb can be made of sufficient strength with less weight and more durable when made of the materials heretofore employed."

With her retinal implant surgery, Dilek Umrancan now sees light and shapes.

Bionic Eye

Patent Name: Retinal Prosthesis and Method of Manufacturing a Retinal Prosthesis
Patent Number: US 8527057 B2
Patent Date: September 3, 2013
Inventors: Robert J. Greenberg, Jerry Ok, Jordan Neysmith, Kevin Wilkins, Neil Hamilton Talbot, and Da-Yu Chang, of Sylmar, California

What It Does   This retinal prosthesis, composed of a camera system on a pair of glasses and electronic implant in the eye, restores partial sight to people who have various levels of blindness.

Background   The first experiment to restore sight with an implantable device was conducted in 1968, when G. S. Brindley and W. S. Lewin surgically implanted a device in a fifty-two-year-old blind woman. They didn't put it in the patient's eye, however, but rather in the occipital lobe of her brain. By sending an electric signal through electrodes in contact with neurons in the brain, they succeeded in causing the patient to see points of light in half her field of view.

As electronics got smaller in the twenty-first century, the focus shifted to prostheses that could be implanted directly into the retina. More than half of all cases of blindness are caused by the degeneration of photoreceptor cells in the retina, the result of diseases such as retinitis pigmentosa or age-related degeneration. Many private companies are working to develop a retinal prosthesis, sometimes called a bionic eye, that could restore partial vision to people who have suffered retinal degeneration.

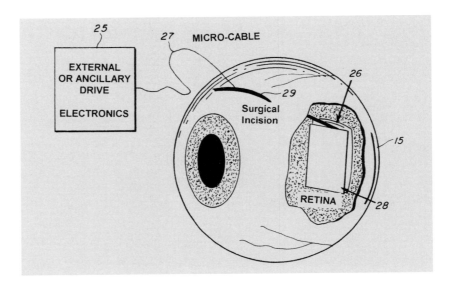

In February 2013, Second Sight's Argus II retinal prosthesis became the first retinal implant to gain FDA approval for clinical trials in the United States. The Argus II device consists of a camera, transmitter, and video-processing unit on a pair of normal eyeglasses, as well as an implantable membrane containing sixty electrodes. The camera sends signals to the processor, which wirelessly transmits signals to the implant device to trigger the electrodes touching the surviving neurons in the retina.

In February 2017, a Milwaukee-area man who had been blind for twenty-five years due to retinitis pigmentosa received the Argus II prosthesis and regained a semblance of sight. One of dozens of clinical patients, Phil DiMeo gained the ability to recognize basic objects and people, and to navigate his house. The retinal implant causes DiMeo to "see" flashes and blurs of light. He does not see in the traditional sense, as the Argus II allows him to perceive images and movement even with his eyes closed. It takes time and practice to learn to recognize objects with the device, but as the technology improves and more electrodes can stimulate nerves in the retina with more precision, sharper definition and clearer images can be achieved. Researchers believe that the technology will soon advance to the point that blind people will be able to regain the ability to read using a retinal implant.

How It Works

1. The process starts with a small camera mounted on a pair of conventional eyeglasses.
2. The raw image data is then sent via a transmitter to a video-processing device, which can either be mounted on the glasses or worn around the neck.
3. The video processor wirelessly sends commands to the implanted retinal prosthesis, which is a soft electronic membrane with a number of small electrodes.
4. The surgery to implant the prosthesis, which takes roughly four hours, places the device near the retina, at the back of the eye.
5. Electrical signals from the prosthesis stimulate the remaining retinal neurons, or photoreceptors.
6. The patient's brain perceives the electrical signals as patterns of light, from dots to cloudy images, which help him identify objects and movement.

7. With time, the patient's photoreceptors become more attuned to the signals from the implanted device, and the patient receives clearer images and gradually learns to recognize them.

8. The same basic technology can be scaled to include more, smaller electrodes on the implanted membrane, generating a more complete image of the world.

In the Inventors' Words "It was at Johns Hopkins University, in the early 1990s, that I, along with Drs. [Eugene] de Juan, [Mark] Humayun and [Gislin] Dagnelie, with some early support from FFB [Foundation Fighting Blindness], carried out the very first short-term (one hour or so) experiments with human volunteers. We demonstrated that electrical stimulation of the retina could, indeed, produce spots of light perceived by blind patients." —Robert J. Greenberg

A 1920's Sears ad for bras, where the most expensive was $1.60

Bra

Patent Name: Brassière
Patent Number: 1,115,674
Patent Date: November 3, 1914
Inventor: Mary P. Jacob, of Mamaroneck, New York

What It Does Confines, shapes, and supports the bust comfortably, with a minimum of fabric—especially in the back—to facilitate its use under formfitting, revealing, or low-backed garments.

Background

In 1913, Mary Phelps Jacob said good-bye, once and for all, to the uncomfortable corsets of the day. These undergarments were made with steel rods and whale bones. They did not conform naturally to a woman's body, nor were they well adapted to shifting fashion trends. While she was getting ready for an evening on the town, the inventor gazed at her own image and saw that the beauty of her brand-new dress was badly compromised by the poking framework of the underlying corset. So instead of wearing the corset, she took a couple of silk handkerchiefs and some ribbon and made herself a substitute. Though it did not consist of the double-cup design that would ultimately become the prototype, Jacob's resourceful innovation would soon provide the basis for the first widely accepted contemporary brassiere.

How It Works

It was the Russian immigrant Ida Rosenthal who founded Maidenform in 1928 and categorized bras into cup sizes.

Victoria's Secret's "Red Hot Fantasy Bra" has a place in the Guinness World Records as the world's most expensive bra. It contains more than 1,300 precious stones, including Thai rubies and diamonds. While it may enhance the bust, it's sure to bust the bank: The bra was priced at $15 million.

1. In figure 1, two pieces of angle-cut cloth (10) are joined together by a seam (13) that extends vertically across the middle.
2. The top edges of the cloths are trimmed shorter than the bottom edges.
3. Elastic shoulder straps (16) are sewn onto the top and bottom outer edges of the two cloths.
4. Two ties or strings are sewn into bindings from the top outer edges of each cloth to the bottom, with a significant length (15) remaining.
5. The angle at which the two ribbons are sewn allows for a low fastening behind the wearer's back.

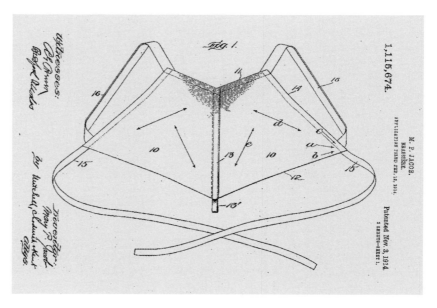

In the Inventor's Words

"The garments, in order to serve their purpose with low corsets, require to be snug fitting in order to shape the figure properly above the waist to confine the bust and conceal the corset top. Garments hitherto proposed for this purpose have required lacing or other fastening of parts across the back of the wearer or have been otherwise arranged so high as to interfere with the wearing of evening gowns cut low in the back. The necessity of a snug fit has also made it necessary that the brassiere be made with a special regard to the size and shape of the wearer in order to obtain a proper fit."

Man in an iBot 3000 at the Metropolitan Museum of Art in New York City

iBot

Patent Name: System and Method for Stair Climbing in a Cluster-Wheel Vehicle
Patent Number: 6,311,794
Patent Date: November 6, 2001
Inventors: John B. Morell, of Manchester, New Hampshire; John M. Kerwin, of Weston, Massachusetts; Dean L. Kamen, of Bedford, New Hampshire; Robert R. Ambrogi, of Manchester, New Hampshire; Robert J. Duggan, of Strafford, New Hampshire; Richard K. Heinzman, of Francestown, New Hampshire; and Brian R. Key, of Pelham, New Hampshire; assigned by DEKA Products Limited Partnership

What It Does   This sophisticated version of a wheelchair is intricately balanced and powered so that it can go where no wheelchair had gone before—up and down stairs, over curbs—virtually anywhere a mobile person can go. Its easy "joystick" operation also allows for eye-to-eye contact with the world.

Background

A significant departure from all previous wheelchair innovations, the iBot carried a considerable price tag, selling at around $24,000 per unit. In 2009 the iBot was discontinued because of its exorbitant price, which was not covered by insurance. In 2016 Toyota and DEKA partnered up to announce that they would be releasing an updated and cost-effective device.

Dean Kamen's name has become inextricably linked with his much-publicized invention, the Segway scooter. That product may not have caught on quite as readily as his supporters believed it would, but Kamen will be remembered for many other significant and humane patents as well. Before inventing the Segway, Kamen was immersed in improving medical technologies and has a long list of credits and patents to his name. After establishing the DEKA Research & Development Corporation in Manchester, New Hampshire, Kamen and a team of engineers developed the iBot, which employs many of the same technological principles that later were used for the Segway.

With two pairs of midsize wheels on a swivel and a complex set of computer-controlled gyroscopes that continually self-balance, the iBot is an extremely sophisticated machine that offers people with paralysis an incredible alternative to traditional wheelchairs, and nearly as much freedom to move around as an able-bodied person enjoys. Controlled by a joystick, it can climb curbs and go up and down stairs, and can be raised to give chair-bound people an eye-to-eye view with someone standing tall. Kamen introduced his product through Independence Technology, a subsidiary company of Johnson & Johnson.

How It Works

The iBot employs wheel and cluster control laws that relate to the location of the center of gravity. A computer constantly monitors the angle at which the vehicle travels. The front and rear angles of the wheels are constantly updated, so that during stair-climbing, for example, the rider can vary the position of the vehicle's center of gravity with very little effort. As a safety feature, a brake pitch control monitors the cluster and wheel motor temperature during stair-climbing. If temperatures exceed a preset value, the brake pitch controller turns off the motor amplifiers and controls the motion by modulating the cluster brakes. The controller places the vehicle in a configuration where all four wheels are on the stairs, thus placing the vehicle in a statically stable configuration with respect to gravity and preventing the rider from continuing in stair mode. If the motors cool sufficiently, iBot will resume stair mode in order to allow the rider to complete the ascent or descent.

1. The operator's seat (12) is attached to a ground-contacting module (16), which contains a power source, drive amplifiers, drive motors, and a controller.

2. These components drive a cluster (36) of wheels (18).

3. The wheels are mounted on a cluster arm (40), and each wheel is capable of being driven by a controller (8).

4. The cluster arm rotates on an axis (22), the rotation of the arm governed by the controller.

5. The controller comprises a processor in communication with a memory storage device, which executes a control program.

6. A number of sensors (38) detects the precise state of the vehicle and receives commands from the rider's control stick (44).

7. These sensors can detect the pitch, roll, and yaw of the vehicle (10) as well as such variables as the angular position and/or rate of rotation of the wheels and cluster arm.

8. The independent control of the clusters and wheels allows the vehicle to operate in several modes, thereby allowing the rider or the controller to switch between modes in response to the local terrain.

FIG. 1

In the Inventors' Words "The device has a plurality of wheels rotatable about axes that are fixed with respect to a cluster arm, where the cluster arm itself is rotated about an axis so that wheels rest on successive stairs. The wheels and cluster arms are controlled according to separate control laws by a controller. Whether the device ascends or descends the stairs is governed by the pitch of the device relative to specified front and rear angles. Shifting the center of gravity of the combination of device and payload governs the direction of motion of the device."

Dean Kamen, prolific inventor

Dean Kamen, an inventor, engineer, and physicist, has a passion for science and technology, and high hopes for what they can accomplish. Born in 1951 on Long Island, he attended Worcester Polytechnic Institute. While still in school, Kamen made his first significant contribution to medical technology and to society. He invented a wearable device that administered small, precise amounts of medication over time. His innovation allowed patients who relied on consistent or periodic injections of medication to move about freely without being tied down to a machine. The idea changed the lives of thousands who had previously been immobile.

On the strength of this invention, Kamen went on to found DEKA Research & Development Corporation, a laboratory that incubated many more ideas related to medical research and technology. One of the lab's more notable inventions is a portable dialysis machine that allows people with kidney disease to receive dialysis at home.

Kamen's inventive energy often focuses on ways to make life easier for people around the world. The Segway is a two-wheeled, upright vehicle with an electric motor that can go 12 miles an hour. Kamen hopes the Segway will become an alternative to more polluting forms of transportation, like automobiles. He envisions a world in which people in cities will use Segways instead of cars to get around, causing levels of pollution and congestion to go down—though so far, they have been slow to catch on in that capacity.

Kamen is a visionary, and he puts a lot of faith in young people. He created an interactive learning center called Science Enrichment Encounters, in Manchester, New Hampshire, and gives TED talks. To help inspire and encourage young scientists, he also founded the nonprofit organization FIRST (For Inspiration and Recognition of Science and Technology) in 1989. FIRST conducts a national robot-building competition and offers over $1 million in scholarships to budding scientists and engineers. Kamen firmly believes that technology can be harnessed to solve many of the world's problems, like famine, pollution, the destruction of the ozone layer, and the contamination of the environment. He is currently working on a device that purifies drinking water, which will help eliminate waterborne diseases. He hopes to distribute the final product to poor families in developing countries. With his dedication to kids and the environment, and his infectious passion for science, Kamen has been the recipient of numerous awards and honors. He continues to dream up inventions designed to improve the quality of life on Earth. He holds more than 440 U.S. and foreign patents, many for innovative medical devices, and in 2005 he was inducted into the National Inventors Hall of Fame.

Kevlar

Patent Name: Wholly Aromatic Carbocyclic Polycarbonamide Fiber Having Orientation of Less Than About 45°
Patent Number: 3,819,587
Patent Date: June 25, 1974
Inventor: Stephanie Kwolek, of Wilmington, Delaware

What It Does

Kevlar's unique internal fiber structure and exceptionally high tensile properties make it ideal for manufacturing the resilient and lightweight body armor used by law enforcement officials and the military.

Background

There are many other significant uses for Kevlar, including strengthening suspension bridge cables, making bomb-resistant aircraft baggage containers, manufacturing safer and stronger car tires, and even making durable sails for ships.

Stephanie Kwolek was a young chemist employed in 1950 by DuPont's Pioneering Research Laboratory in Wilmington, Delaware. Her experiments during the 1960s resulted in the development of synthetic fibers much stronger than any created before. In fact, the material she invented is five times stronger than its weight in steel. DuPont instantly found a commercial use for Kwolek's new crystalline polymers: Kevlar "bulletproof" vests. Hundreds of lives have been saved annually as a result of this invention.

Kevlar has received a lot of publicity recently in light of world affairs, and the need for bulletproof fabric in the fight against terrorism. Kwolek was proud to have contributed to an invention that has helped preserve human life. Even after retirement, she remained devoted to improving the world through science.

"I can't see how anybody can live in this modern world without an understanding of science," she says, "especially because it touches every aspect of our lives."

How It Works

1. Polyparaphenylene terephthalamide is produced through a chemical process called polymerization.
2. The crystalline liquid with polymers is extruded from a spinneret, a metal plate studded with tiny holes.
3. The fibers that emerge through the holes are cooled in water to allow them to harden.
4. The fibers are subsequently twisted together into yarn that can be woven into a durable synthetic cloth.

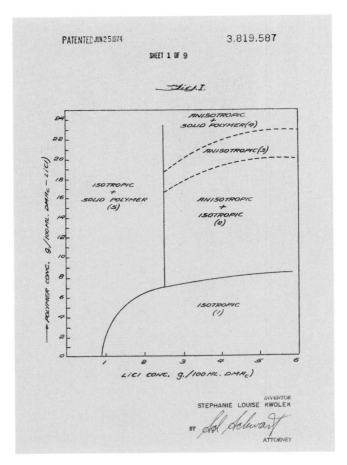

Fig. I.

STEPHANIE LOUISE KWOLEK
INVENTOR

BY *[signature]*
ATTORNEY

In the Inventor's Words

"This invention relates to novel, optically anisotropic dopes [absorbent or adsorbent material] consisting essentially of carbocyclic aromatic polyamides in suitable liquid media. These dopes, and related isotropic dopes, are used to prepare useful fibers, films, fibrids, and coatings. In particular, fibers of unique internal structure and exceptionally high tensile properties are provided."

Laser Eye Surgery

Patent Name: Far Ultraviolet Surgical and Dental Procedures
Patent Number: 4,784,135
Patent Date: November 15, 1988
Inventors: Samuel E. Blum, of White Plains, New York; Rangaswamy Srinivasan, of Ossining, New York; and James J. Wynne, of Mount Kisco, New York

What It Does

Gets rid of your glasses, for one thing. The ultraviolet surgical and dental procedures described in this patent broke new ground with the use of a laser that photoetches organic biological matter without using heat.

Background When he was a doctoral student at New York's Columbia University in the '50s, Gordon Gould first coined the term "laser"—an acronym for "light amplification by stimulated emission of radiation." The word was only partly original: He'd been studying under the professor who invented "maser" (microwave amplification by stimulated emission of radiation). His new term would soon overshadow his professor's. Gould believed that light-wave amplification had a lot more potential, and power, than microwave amplification. It turned out that he was correct.

By the 1980s, laser technology was being used in a variety of fields, including eye surgery. In this field, lasers were only used to create therapeutic scar tissue, until three IBM scientists—the awardees of this patent—began using a short-pulse ultraviolet laser called the excimer laser, which—unlike previous lasers—did not generate substantial heat. It did not burn tissue during surgery and it was also much more precise. It opened a whole range of surgery based on a new ability to improve, rather than simply to repair.

LASIK, which stands for "laser-assisted in situ keratomileusis," is one such example. The excimer laser is used to permanently change the shape of the cornea to correct flawed vision. The surgery has already helped millions of people all over the world to put down their glasses and contact lenses for good. The impact LASIK surgery has had on the traditional corrective lens industry would make an interesting study.

LASIK eye surgery

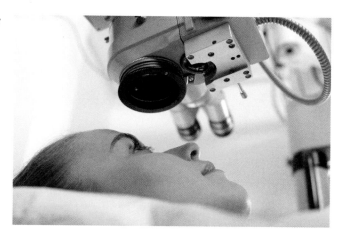

Another application of the invention is in the field of dental medicine, in which the far-ultraviolet radiation of wavelengths can be used to remove decay from teeth without damaging the surrounding enamel.

How It Works
1. The patent illustration shows a source of ultraviolet radiation of wavelengths less than 200 nanometers, such as the excimer laser.
2. A casing is used to contain the laser beam.
3. The casing includes a shutter, which can either block the radiation beam or allow it to pass.
4. A reflecting mirror is used to change the direction of the radiation beam.

5. A lens can be used to focus the radiation beam onto a selected spot of the organic material. An aperture provides more precise alignment of the radiation before it strikes the lens.

6. The instrument is attached to a moveable arm with articulated joints so that a surgeon can move the beam around in precise manner.

In the Inventors' Words

"This invention relates to surgical and dental procedures using far ultraviolet radiation of wavelengths less than 200 nm, and more particularly to a method and apparatus for selectively removing organic material without heating and damage to surrounding organic material."

An MRI machine

Magnetic Resonance Imaging

Patent Name: Apparatus and Method for Detecting Cancer in Tissue
Patent Number: 3,789,832
Patent Date: February 5, 1974
Inventor: Raymond V. Damadian, of Forest Hills, New York

What It Does

Analyzes tissue for presence and degree of cancerous malignancy, serious muscle strain, and other important medical diagnostic information.

Background

The original machine was dubbed "Indomitable," and it is now in residence at the Smithsonian Institution.

The first MRI exam performed on a human being took place on July 3, 1977.

Magnetic resonance imaging (MRI) is a means by which soft tissue, including musculature, is photographed to detect certain structural pathologies and anomalies. The technique relies on the magnetic properties of hydrogen atoms, which are present in most organic molecules and water in the human body. When the hydrogen atoms are immersed in a powerful magnetic field and bombarded with radio waves, they emit radio signals that provide information about their local environment. MRI technology provided doctors with an entirely new and profoundly significant view inside the human body.

MRIs are used in the U.S. for more than one million scans each year. Many people experience the anxiety of claustrophobia at the thought of going into the dreaded MRI tube. A more legitimate fear is of what might fly toward the tube when the magnetic field is turned on. Though relatively safe as a diagnostic machine, the powerful magnetic force of the MRI apparatus has caused accidents, including a tragic one in which an oxygen tank flew forcefully into the magnetic field, killing a six-year-old boy.

The 2003 Nobel Prize in Physiology or Medicine was awarded to Paul C. Lauterbur, of the University of Illinois, and Sir Peter Mansfield, of the University of Nottingham in England, who are acknowledged as MRI's inventors, despite Damadian's previous involvement. The two men made significant improvements upon Damadian's original idea—including the employment of a magnet gradient principle—considered distinct innovations in their own right. Embittered, Damadian placed news ads denouncing the Nobel committee's decision as an oversight. But the Nobel committee has never reversed a decision, and since deliberations are confidential, Damadian will never know if he was ever seriously considered.

One thing is certain: His work was crucial to the development of the technology used today. In the early 1970s, Damadian reported in the *Journal of Science* that the radio signals emitted by cancerous tissue were detectable as markedly different from those of healthy tissue. He envisioned the day when MRI, then called nuclear magnetic resonance, would be able to detect these signals. He drew blueprints for such a machine and received several patents, including this one. Damadian says that he essentially laid the groundwork for the discovery—and he is not alone. In 1989, Damadian was inducted into the National Inventors Hall of Fame, which cites him as the creator of "the magnetic resonance imaging (MRI) scanner, which has revolutionized the field of diagnostic medicine."

How It Works   The MRI scanner has three basic functions: magnetizing, applying radio waves, and using imaging technology to draw pictures. First, a part of the body in question, or the entire human body, is subjected to powerful magnetic fields. Inside the human body, millions of tiny atoms—which happen to have magnetic properties—line up with the magnetic field created by the machine. Secondly, radio waves are applied into the field to tip the atoms out of line with the field, simultaneously imposing energy into them. When the radio waves are removed from the field, the atoms reposition themselves using the energy imparted by the radio waves. In doing so, they emit subtle radio signals that are picked up by a highly receptive antenna. Deformities can be detected by fluctuations in the radio signals. Lastly, a computer registers the radio signals and produces a black-and-white image.

1. In figure 2, an electromagnet (23) is displayed in cross-section.
2. A transmitter probe is fixed on a track (72) that can slide along the housing in a helical pattern for access to various body parts.
3. The transmitter probe includes a beam-focusing mechanism (71) that concentrates magnetic energy from a radio frequency generator.
4. A similarly arranged receiver probe (73) detects and transmits the magnetic energy.
5. The induction apparatus registers images based on the view obtained by the magnetic probes.

PATENTED FEB 5 1974   3,789,832
SHEET 2 OF 2

NUCLEAR INDUCTION
APPARATUS & DISPLAY

FIG. 2

FIG. 3

In the Inventor's Words

"This invention relates to an apparatus and method of detecting cancer in tissue and, more particularly, to the use of nuclear magnetic resonance techniques to detect alterations in the organization and structure of selected nuclei in the tissue which alterations are believed to be caused by cancer."

Motorized Exoskeleton

Patent Name: Locomotion Assisting Device and Method
Patent Number: US 8905955 B2
Patent Date: December 9, 2014
Inventors: Amit Goffer and Chaya Zilberstein, of Yokneam, Israel

What It Does

This wearable mobile machine uses electric motors, pneumatic pumps, hydraulics, and other mechanical technologies to allow a person with limited mobility to walk and move or to provide additional strength and endurance to the user.

Background

The first device that could be called an exoskeleton was the "apparatus for facilitating walking" developed by Russian inventor Nicholas Yagn in 1890. The wearable machine used bags of compressed gas to store energy and assist with movement, though it had no external power source and required human power to use.

The first motorized exoskeleton came in the 1960s, when General Electric and the U.S. military built a mobile machine suit called Hardiman. The bulky suit was designed to amplify the strength of the user by a factor of twenty-five, with the goal of letting the wearer lift loads of up to 1,500 pounds. Powering the full exoskeleton resulted in violent and uncontrolled movement, however, and the suit was never tested with a person inside.

The military continued to invest in exoskeleton research both for prosthetic aids and with the goal of designing a suit that could improve the strength and endurance of soldiers. As humanoid robotics technologies advanced, private companies also began investing in research and design projects. In 2014, the FDA allowed marketing of ReWalk, the first exoskeleton intended to allow people with lower-limb paralysis from spinal cord injury to walk and move again.

The ReWalk device was used in rehabilitation facilities before being approved for home use. It is designed to help users sit, stand, walk, and climb stairs. The mobile machine consists of leg braces, motorized joints and actuators, motion sensors, a wireless remote control, and a central computer processor and battery worn as a backpack.

Additional exoskeleton suits are being developed as well, such as the EksoWorks unpowered exoskeleton, which uses counterweights to allow workers to hold power tools and other heavy objects for long periods of time without fatigue. Powered exoskeletons developed by engineers at institutions like MIT and the European Space Agency continue to be refined. Although significant physical therapy and assistance is still required to teach people with paralysis to walk again, the suits continue to get lighter and more precise. Future exoskeletons will likely be cheaper and more widely available to people with mobility impairments, providing an alternative to wheelchairs.

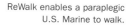
ReWalk enables a paraplegic U.S. Marine to walk.

How It Works

1. A powered exoskeleton requires a battery and a computer system that runs algorithms to interpret data from sensors and controllers on the suit. These two components are generally worn in a backpack while using the suit.
2. The suit includes a central support structure, such as a trunk, and leg braces with electric motors and actuators to move the user's limbs.
3. A wireless controller allows the wearer to input commands, such as, while seated in a chair, telling the suit to stand up.
4. In addition to the commands provided by the user, the suit monitors a series of sensors on the joints to determine the position of the wearer.
5. Algorithms in the computing system make adjustments to the machine's movements based on the position of the user's body.
6. Sensors in the bottoms of the wearer's feet monitor the pressure against the ground, letting the exoskeleton assist with walking movements while remaining balanced.

In the Inventors' Words

"ReWalk has been my life's work and passion from well before I founded the company. The technology and its impact on the daily lives of individuals with spinal cord injuries have exceeded my expectations. This is just the beginning of the journey for individuals with lower limb paralysis and those with higher order injuries." —ReWalk co-founder Amit Goffer

Nicotine Patch

Patent Name: Transcutaneous Application of Nicotine
Patent Number: 4,597,961
Patent Date: July 1, 1986
Inventor: Frank Etscorn, Socorro, New Mexico

"He lit the cigarette and smoked it down to the filter in one breath. He silently thanked the cigarette company for being thoughtful enough about his health to include a filter to protect him. So he lit up another." —Steve Martin, "The Smokers"

What It Does

Designed to ease the withdrawal process when quitting smoking, the patch provides a steady, controlled release of nicotine through the skin, without the tars and toxins of actual cigarettes.

Background

Studies reveal that nicotine itself actually has medicinal properties known to have positive effects in the treatment of attention deficit disorder, Parkinson's disease, Alzheimer's disease, Tourette's syndrome, ulcerative colitis, and schizophrenia.

The Outlaw of Marlboro Country: Both Nicotrol and Nicoderm are the commercial products of co-inventor Jed Rose, who is often credited as the inventor of the patch. Rose is a chief researcher at Duke University, located in Durham, North Carolina, near many major cigarette companies.

Sipping a Cig? Among the most interesting smoking cessation aids developed, a one-time-use straw delivers tiny beads of nicotine while its user sips a beverage of his choosing. Presumably, there is a psychological correlation to the physical similarities between straws and cigarettes.

After receiving a doctoral degree in experimental psychology from George Peabody College in Nashville in the mid-1970s, Frank Etscorn moved to the Southwest to teach at the New Mexico Institute of Mining and Technology. While there, he used a favorite ingredient of local fare—the spicy capsicum or chili pepper—to illustrate a theory suggesting that not only are humans able to turn painful experiences into pleasurable ones, they can develop cravings for the natural chemicals that the body releases in the process. "We need a fix of red or green chilé with a side order of endorphins," he reported in 1990 in the *Albuquerque Journal.*

Others have expanded on this pain-to-pleasure principle to uncover the nature of addiction and reveal how our species is perhaps the only one that willfully ingests what are otherwise intolerable substances, such as nicotine. Dr. Etscorn had the serendipitous experience of ingesting the drug—not willingly, by accident!—when he spilled some liquid nicotine in his lab in 1979. He could feel the substance being absorbed through his skin, or "transcutaneously," and this led to his invention of the nicotine patch.

While science has shed great light on the hazards of smoking, the hard-to-kick habit is still a leading cause of disease and death throughout the world. In 1988, the U.S. Surgeon General's Office released a report decisively stating that tobacco products containing nicotine are, indeed, addictive. By the mid-1990s, smoking cessation was an industry in itself. Gums, mints, and other nicotine substitutes or supplements, including the patch, were all aggressively marketed. The patch was considered by many experts—and consumers—the most effective. There are several major brands on the market, including Nicoderm, Prostep, and Habitrol, the one Etscorn's invention established. Nixing its prescription requirements, in 1996, the U.S. Food and Drug Administration approved over-the-counter sale of nicotine patches.

How It Works
A measured dosage of nicotine is administered through a tiny reservoir fixed into an absorbent adhesive pad that is applied to the skin. Dosages are lowered over time, gradually diminishing nicotine dependency until it is gone. In the illustration, a nicotine-infused bandage (10) has a diameter of about 1.5 centimeters and a thickness of about 2.0 centimeters. A nicotine impermeable backing (12) made of impervious plastic or rubber material contains a cavity (14) along one surface to contain the liquid nicotine (16). Sealing the nicotine film (18) again is a suitable flexible material inert to the nicotine. Adhesive (20) is provided around the nicotine-permeable membrane.

In the Inventor's Words
"Nicotine appears to be the most pharmacologically active substance in tobacco smoke, yet it appears to be not as significant from a health standpoint as the tars and carbon monoxide. However, nicotine is very important from another standpoint, i.e., it is the reinforcing substance in tobacco which maintains the addiction. In this respect, a theme commonly heard among workers in the field of smoking research is, 'People would be disinclined to smoke cigarettes if an alternate route of nicotine delivery could be devised.'

"Several such attempts have been made to administer nicotine in alternate ways, but with varying and generally ineffective results. For example, nicotine-containing pills have been studied; however, effective blood levels of nicotine are not achieved. Drugs absorbed in the stomach pass through the liver first where, in this case, 80–90 percent of nicotine deactivation occurs. Similar findings have been demonstrated with nicotine chewing gum although it has been sufficiently successful to warrant its marketing."

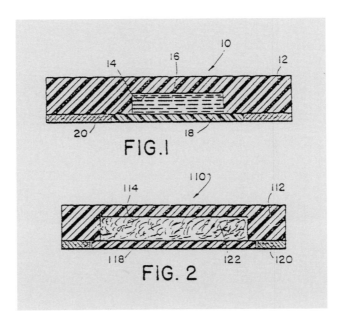

Parachute

Patent Name: Life-Saving Device for Aeroplane
Patent Number: 1,332,143
Patent Date: February 24, 1920
Inventor: Karl O. K. Osterday, of Cincinnati, Ohio

What It Does

Provides a safe, low-impact means of descending from great heights (as in exiting an airplane while in flight) by deployment of a fabric canopy that uses wind resistance to slow the pace of descent and reduce the force of impact.

Background

Parachuting or "skydiving," in a variety of forms, is a popular recreation for adrenaline junkies across the world. On October 24, 2014, Google's Alan Eustace achieved the highest parachute jump, plunging 135,890 feet in a specialized space suit to set the record.

The idea of floating from a great height to land safely on terra firma has long occupied humans. Leonardo da Vinci sketched a version of a parachute some five hundred years ago, with the idea of using it to rescue people from burning buildings. Since then, many have made and employed primitive versions of parachutes, including a Frenchman who dropped a dog to safety in a chute following a hot-air balloon mishap. World War I expedited the development of improved parachutes that could be easily contained and quickly employed in emergency situations involving warplanes. Karl Osterday was by no means the father of the parachute—in fact, his invention was not a favorite among aviators, who preferred detached packs rather than those worn on the back. Nonetheless, some of the principles Osterday developed for this patent led directly to subsequent improvements on the parachute, including multiple canopies and a mechanism to allow the user to free himself of the parachute upon landing. It is a device that has saved countless lives in both war and peace.

How It Works

1. Belts and straps are fastened around an aviator's waist, torso, and thighs, forming a secure harness.

2. The straps connect at the back to a canvas pack containing two of three parachute canopies.

3. A third canopy is contained in a smaller pack mounted on the airplane itself. A slipknot is fastened to a ring on top of the aviator's helmet, and is attached to the aviator's wrist by means of a cord.

4. If a pilot loses control of the airplane, he pulls the cord (29) so as to open the receptacle (27), and then quickly jumps out of the plane.

5. Canopies (21, 22) are dragged out of the receptacle by cords (24) and immediately fill with air.

6. These canopies also engage the cords to activate the other canopy.

7. As the pilot begins his landing, he can disengage himself from the entire parachute by pulling on a cord (18).

In the Inventor's Words

"One object of the present invention is to provide a novel and improved parachute device by means of which the aviator can release himself from his wrecked machine, while in the air, and make a safe landing.

"Another object is to provide a novel and improved device of this character wherein the aviator can release himself from the parachute when within a desired and safe distance from the ground, so that he will be unencumbered upon landing."

Prozac

Patent Name: Aryloxyphenylpropylamines
Patent Number: 4,314,081
Patent Date: February 2, 1982
Inventor: Bryan B. Molloy, of Indianapolis, Indiana, and Klaus K. Schmiegel, of Indianapolis, Indiana; assigned by Eli Lilly and Company

What It Does

This psychotropic medication, taken in pill form, has proved a godsend for many sufferers of chronic depression by inhibiting reuptake of serotonin in the brain, and thereby altering mood and state of mind without causing hallucinations or other harsh side effects.

Background

While selective serotonin reuptake inhibitors such as Paxil, Prozac, Luvox, Effexor, Serzone, Anafranil, and Zoloft have enjoyed a great deal of success in treatment, they also have their share of detractors. Many books have been written about their downsides, and about their implication in society and natural evolution.

Prozac was first introduced in the U.S. in 1988. In 2010, more than 24.4 million prescriptions for the drug, and its generic form, were filled.

In his memoir *Darkness Visible*, novelist William Styron recounts his own personal struggle with a condition that nearly claimed his life. Indeed, depression is a very serious medical condition, too often with fatal consequences. While skeptics persist in disparaging what they do not know, neuroscience, psychological research, and gifted, insightful writers such as Styron are helping move the condition from beneath the social shadows of stigma to shed light on the faulty brain chemistry that plagues countless numbers of people. The fact that there may be a large population who will jump at the chance to feel better (whether they really need to or not) should not be a negative reflection on those who genuinely benefit from the help that antidepressants offer; nor should it discourage those who feel they need such help. Enough stigma remains.

One revolutionary breakthrough in the treatment of depression came with the discovery of selective serotonin reuptake inhibitors (SSRIs). Serotonin is a chemical messenger secreted by one nerve cell and picked up by another. It is believed that many people suffer the symptoms of depression because serotonin is absorbed too quickly. Prozac became the first in a growing class of commercial antidepressant medications known as SSRIs, which are believed to increase

Depression does not discrimi-
nate; it afflicts all kinds of people
from all walks of life. American
Psychological Association statistics
show that approximately 14.8
million American adults have major
depressive disorder, and it is the
leading cause of disability in the
U.S for ages 15–44.

serotonin levels in the brain. "Aryloxyphenylpropylamines" include the active
ingredient in Prozac, the compound known as fluoxetine hydrochloride.

In the past, antidepressant medications addressed dopamine and
norepinephrine, other compounds in the monoamines group. Along with
these, serotonin plays significant roles in mood regulation, eating behavior,
aggression, and pain. The success of antidepressants has helped to prove
that depression is caused by an imbalance of neurotransmitters in the brain.

How It Works

In the limbic area of the brain, serotonin is blocked from reabsorption on
the synapses between cells. However, brain chemistry is still an extremely
mysterious and largely unknown frontier. Little is known about exactly
how in balancing the neurotransmitter levels it achieves results for patients.
Experiment and research cautiously continue in the study and development
of the field. Doctors will often gradually increase doses to judge the success
of the drug and to minimize side effects.

In the Inventors' Words

"The tricyclic antidepressant drugs presently being marketed inhibit the uptake of monoamines by brain neurons, most of them being more effective
in inhibiting the uptake of norepinephrine than of serotonin. Many of the
compounds of this invention behave similarly in that they block norepinephrine uptake more effectively than they do serotonin uptake. The
p-trifluoromethyl derivative, however, is far more effective in inhibiting serotonin uptake than in inhibiting norepinephrine uptake. Thus, although
the compounds of this invention clearly have potential as anti-depressant
compounds, it is apparent that N-methyl 3-(p-trifluoromethylpheny-
loxy)-3-phenylpropylamine and its tertiary and primary amine analogs
will have a different type of anti-depressant action from the presently
marketed drugs."

Reading System for the Blind

Patent Name: Reading System
Patent Number: 6,199,042
Patent Date: March 6, 2001
Inventor: Raymond C. Kurzweil, of Newton, Massachusetts

What It Does

Translates written words into audible speech.

Background

Invented in 1976, the Kurzweil Reading Machine was the world's first computer to
transform text into computer-spoken words. It scans print and uses technology
that recognizes letter patterns to "read aloud" the material on the page, thus
providing the blind and visually impaired access to virtually anything in print.
The Kurzweil Reading Machine is considered the most significant advance for
the blind since Braille was developed almost 150 years prior.

The system is an example of the enormous benefits that new technologies (computer chips and optical recognition software) can yield. And it is only one of the creations on this patentee's impressive resume. In addition to the first print-to-speech reading machine for the blind, Kurzweil helped to develop the first omni-font optical character recognition method, the first CCD flatbed scanner, a text-to-speech synthesizer, and the first commercially marketed large-vocabulary speech recognition system.

How It Works

Stevie Wonder nominated Kurzweil for the nation's most prestigious award in invention and innovation, the $500,000 Lemelson-MIT Prize, which Kurzweil received. The inventor and the famous musician-composer forged a friendship that inspired another Kurzweil invention: a system of computer-based instruments that could realistically reproduce the sounds of acoustic instruments, such as the grand piano.

1. The reading system includes a computer, a mass-storage device, and software that allows the machine to "read" an optically scanned image of a document.
2. The software converts image files into text files, and relays positional information associating the text with its image representation.
3. In synchronization with a highlighting indicator applied to a displayed representation of a document, the computer selects from a series of prerecorded voice samples.
4. While highlighting is applied to the words on the monitor, the reading machine plays back the stored, recorded voice samples.

In the Inventor's Words

Kurzweil was inducted into the Inventors Hall of Fame in 2002.

Raymond C. Kurzweil

"In one aspect of the invention, a computer program product residing on a computer readable medium includes instructions for causing a computer to display a representation of a document on a computer monitor. The product also causes the computer to read the displayed representation of the document by using a recorded human voice. Optionally the recorded human voice is synchronized with highlighting applied to the displayed representation. The computer program product uses information associated with a text file to synchronize the recorded human voice and the highlighting to the displayed representation of the document."

Man in an Iron lung

Respirator

Patent Name: Respirator
Patent Number: 3,191,596
Patent Date: June 29, 1965
Inventor: Forrest M. Bird, of Palm Springs, California, and Henry L. Pohndorf, of El Cerrito, California

What It Does   This pulmonary aid assists people with breathing problems by duplicating the functions of the human respiratory system.

Background   Before the modern respirator, there was the iron lung, a behemoth medical device that looked vaguely like a prehistoric whole-body MRI scanner. A patient would lie inside of the machine to receive treatment for respiratory paralysis, an often fatal symptom of poliomyelitis. (Rare today, polio was rampant before Jonas Salk developed a vaccine to combat it in 1952.) Encased in the chamber, the patient's entire body would be subjected to the pressure of pump-activated bellows that would "breathe" air into the chamber, activating subatmospheric pressure on the patient's chest and, in turn, enabling the patient to breathe.

These machines went the way of the dinosaur after Forrest Bird's innovative spirit swept through. Bird invented the first mass-produced medical artificial ventilation unit: the Bird Mark 7. Forrest Bird held several patents, three of which pertained to the revolutionary nature of his respirator: this one; patent number 3,068,856 for a "Fluid Control Device"; and patent number 3,842,828 for a "Pediatric Ventilator." The last one, which significantly reduced respiration-related infant mortality, was respectfully nicknamed, and trademarked, the Baby Bird.

Another pun on his name could be made in connection with his early career as a jet and helicopter pilot during World War II. Airplanes were then ascending to elevations never attained before, and Bird became interested in the technology used to help deliver oxygen to pilots at high elevation. Bird then steered his career toward medicine and cardiopulmonary physiology, establishing a company in 1954 and going on to develop some of the most successful respirators known.

How It Works Bird's respirator is a complex machine. More than 350 parts are numbered in eighteen separate patent drawings. Yet, while it served a number of highly sophisticated functions, it weighed only about six pounds and had the dimensions of a small shoebox, making it an attractive alternative to the gargantuan iron lung.

1. During its operative cycle, the device works in two phases: an inhalation phase and an exhalation phase.
2. A controller attaches by an inlet to a source of pressurized gas and an outlet to connect to a patient's airway.
3. A main valve controls the flow of gas from inlet to outlet.
4. It remains open during the inhalation phase and closes during exhalation by means of a pressure-sensing mechanism for a predetermined duration.

"Mechanical ventilators impose their design limitations on the patient, some respirators not even offering simple adjustments for combating bronchiolar collapse, and impeding or encouraging venous return. An important object of this invention is to provide a respirator that always provides peak pulmonary conformance, so that the physician, freed from having to devote attention to this problem, can devote his attention to considering respiratory patterns for good tidal exchange, even distribution of alveolar air, facilitation of bronchial drainage, and other adjustments."

Seat Belt

Patent Name: Safety Belt
Patent Number: 312,085
Patent Date: February 10, 1885
Inventor: Edward J. Claghorn, of New York, NY

What It Does

Provides a safety restraint mechanism to keep an individual in any moveable seat. It was later adapted for the seats of powered vehicles such as automobiles and airplanes, to counteract the forward momentum caused upon impact or in case of turbulence.

Background

The first safety belt was devised more than 130 years ago, primarily for work-men who needed to ascend or descend tall buildings. This was before the rise of the automobile, and in retrospect, the invention seems visionary. The use of safety belts in cars and airplanes has saved countless lives and is required by law in many places throughout the world—but its original purpose was more general. According to the patent application, the safety belt was "intended for the use of tourists, sailors, painters, farmers, firemen, telegraph-men, and others."

How It Works

Nils
Bohlin

A Swedish safety engineer named Nils Bohlin invented what is recognized now as the modern safety belt—the three-point seat belt. Not surprisingly, he was working for Volvo, which prides itself on its strong reputation for safety. Volvo was the first automobile company to make safety belts a regular feature of its cars, in 1959.

Restrains motion and locks one securely into a stationary position.

1. A leather outer belt (A) of sufficient length includes at one end a buckle (b) and fastens around a person's waist through typical belt-hole perforations.

2. The outer belt is attached to a wider, protective inner belt (B), which extends nearly throughout the length of the outer belt, where it is firmly sewn or otherwise attached with sufficient open space between the belts.

3. Bands (C, C1, and C2) are fastened to the outer belts and include rollers (d) on their backs to allow them sliding movement along the outer belt.

4. An array of useful items—straps (g) or slotted pockets (D)—can be attached to the bands.

5. Also, a slide (C2) can provide solid housing for a ring (k) that is attached to a spring snap-hook (D2).

6. The opposite side of the snap-hook features a rope-clamping jaw (s) that aids in sustaining the suspension of a person at varying altitudes.

In the Inventor's Words

"This invention relates to belts designed to be applied to the person, and provided with hooks and other attachments for securing the person to a fixed object, as also for making an interrupted or uninterrupted descent there from, and for carrying tools and other applications, so as to leave the hands and feet of the user at liberty when working at an elevation or during his ascent and descent."

Smoke Detector

Patent Name: Smoke Detector
Patent Number: 3,460,124
Patent Date: August 5, 1969
Inventors: Randolph J. Smith, of Anaheim, California, and Kenneth R. House, of Norwalk, California

What It Does

The battery-powered smoke detector is an important in-home safety feature, often required by law, which emits a piercing alarm in the presence of smoke, alerting residents of a possible fire.

Background

Smoke detectors save lives. According to the National Fire Protection Association, homes equipped with smoke alarms have 40–50 percent fewer deaths than homes without them.

Tips for effective use of a smoke detector

- Install at least one on each level of your home.
- Never remove a good battery or otherwise disable the detectors.
- Check batteries often, and replace when necessary.
- Plan a home escape route in the event of a fire.
- Hold practice drills with your family.

The benefits of a residential fire-alert system were quite obvious when the two inventors from the Interstate Engineering Corporation received the first battery-powered smoke detector. The prototype used electricity from an outlet, and batteries provided a backup source of power.

Today, there are two common types of smoke detectors available. One, called an "ion chamber smoke detector," uses a small amount of radioactive material to detect the presence of smoke or heat. This type is more popular, as it is less expensive and more sensitive. The other type, described in this patent, uses a photoelectric sensor to detect the changes in light levels caused by smoke particles.

How It Works

1. In figures 1, 2, and 3, a housing (H) includes a main body (6) sitting within a circular wall (14) with apertures (18) in it.
2. A flat circular portion (20) includes an inward-extending socket (22) which supports a lens (24).
3. The lens transmits light from a light source (26) contained within a casing (28).

4. Light travels through the lens, through an inner aperture (32), and into a diffusion and reflection chamber (34).
5. When the chamber becomes filled with smoke, the diffusion of light is disrupted and light is reflected from smoke particles onto a photoelectric cell (50).
6. When this happens, the cell activates an audible alarm device (76) that is connected through a relay.
7. When the alarm is inactive, a ruby glass indicator (64) on the top of the external face of the housing indicates that the light source is working, and the smoke detector is ready and functional.

In the Inventors' Words   "We have provided a smoke detector which is compact in construction and can be mounted conveniently on the wall of a home or other building structure and which is rendered operative merely by placing a battery in the housing and plugging the leads into the transformer (66) in a conventional wall outlet. In its operative condition, light from the lamp (26) is always present to indicate that condition, but the light is not diffused and reflected to the triggering photoelectric cell until there is smoke in the diffusion and reflection chamber (34). The device is extremely reliable. Its operative condition is indicated by illumination of the ruby glass indicator (64) and the alarm being operative either by current from the conventional 115v. service in the home or by the self-contained battery in the alarm itself."

Turtle Excluder Device

Patent Name: Turtle Excluder Device (TED)
Patent Number: 4,739,574
Patent Date: April 26, 1988
Inventor: Noah J. Saunders, of Biloxi, Mississippi

What It Does   Saves sea turtles caught in nets during the shrimping process.

Truly ancient mariners, sea turtles are among the Earth's oldest species, descended from larger ancestors that predate dinosaurs by some 50 million years. Today, the Kemp's ridley is the most endangered of all sea turtles. But it appears to be in the earliest stages of recovery due to two major factors: protection of nesting females and nests in Mexico, and the requirement of TEDs in shrimp trawls in both the United States and Mexico. In 1996, the U.S. Court of International Trade ruled that all countries that export shrimp to the United States must have an enforcement policy regarding TED usage.

FIG. 5

FIG. 6

Background

The Trash and Turtle Excludes Device, an improvement on the TED, was created in 2009 and reduces accidental capture of larger marine species by up to 40 percent.

Little shrimp mean big business—but sometimes shrimping can be bad business for the environment. Shrimp trawling uses a large net, which sweeps a broad swath of ocean space. Often, the net captures more than the intended quarry, including endangered marine life like the Kemp's ridley sea turtle. Today, United States environmental policy requires the use of a turtle excluder device (TED) when trawling for shrimp. It is believed that a Georgia fisherman named Sinkey Boone invented one of the first TEDs, which was subsequently developed further by the National Marine Fisheries Service (NMFS) between 1978 and 1984. With minimal loss of shrimp, TEDs are effective at excluding up to 97 percent of sea turtles from shrimp nets.

How It Works

1. As the vessel sails, the trawl net extends in a broad arc, funneling shrimp toward the "cod" end, or the neck, of the trawl—a large receptacle that is then lifted out of the sea so the shrimp can be unloaded onto the boat.
2. Affixed to the trawl at the neck is a grid of parallel bars positioned in a gradual upward angle so that, while shrimp glide right on through, turtles can be maneuvered toward the roof of the net enclosure.
3. Upon contact, the turtle shell activates a release hatch that sends it out of the net through a trapdoor, which closes behind it. The patentee for this invention has introduced an improvement to previous TEDs, intended to reduce shrimp loss when the trapdoor opens. The trapdoor is specifically designed to conform to the anatomy of a turtle, allowing no more space than necessary for escape.

"It is the purpose of the current invention to show an improved form for a turtle exclusion device by providing, appended to an angled bar turtle exclusion device, a particular form of trap door escape which is shaped so as to adapt to the shell shape of a turtle, providing for a more ready opening and escape of the sea turtle, and is angled, hinged, and pivoted to the overall excluder so as to provide for a positive closing effect, preventing the loss of shrimp when not actually being activated by an escaping turtle."

Vitamins

Patent Name: Process for Obtaining Vitamins
Patent Number: 2,049,988
Patent Date: August 4, 1936
Inventor: Robert R. Williams, Roselle, New Jersey, and Robert E. Waterman, of Orange, New Jersey; assignors to Research Corporation, of New York, New York

What It Does

The process outlined in the patent describes a method of extracting vitamins from their sources to be used as important nutritional supplements.

Background

A patient afflicted with beriberi, a disorder of the nervous system caused by a lack of vitamin B1

The word *beriberi* means "I can't, I can't" in Sinhalese. It is also the name of a disease caused by malnutrition and specifically a deficiency of thiamine, once common in Eastern countries, where milled rice was stripped of the thiamine layer on the outside. The effects of the disease can be excruciating, causing severe nerve damage, inflammation, and even death. Robert Williams likely witnessed victims in the throes of this disease as a boy. He was born in India to missionary parents, and he remained there until he was ten.

Back in the U.S., Williams attended the University of Chicago, earning a master's of science in chemistry. Then he moved to the Philippines to take a job at the Manila Bureau of Science. He spent much of his energies looking for a cure to beriberi. Williams went back to the States during World War I to work for the Bureau of Science in Washington, DC, and later as the chemical director for the Bell Telephone Company in New York.

Throughout his life, Williams did not forget about the malnutrition suffered by the people from his birth country. Eventually, in 1933, he came up with a way to isolate thiamine in crystalline form. He used what is called fuller's earth—a highly adsorbent claylike substance—to extract the vitamin. Two years later, he could synthesize vitamin B_1, the exact vitamin whose deficiency causes beriberi.

His discovery, or rather synthesis, was a medical marvel that helped boost vitamin synthesis and the field of chemically reconstructed vitamins. Williams's accomplishments also contributed to enriching grains in this country to combat the riboflavin deficiency common among impoverished people throughout the world, including in the U.S.

How It Works

1. Water-soluble vitamins are extracted through an adsorbent filter.

2. The resultant solution is treated and preserved by chemical means.

3. Acyl aromatic chloride, aromatic sulphochloride, or lead acetate can be used to purify the solution.

4. The resulting vitamin solution may be concentrated to a small volume by evaporation in vacuo, or further purified.

In the Inventors' Words

In 1936, Merck & Company first introduced thiamine to the commercial market.

A Federal Enrichment Act passed in the U.S. in 1941 required that millers, bakers, and cereal manufacturers restore the nutritive content of whole wheat products by adding iron, thiamine, riboflavin, and niacin lost in milling or processing.

"According to this invention the vitamin is removed from fuller's earth or similar absorbent by extraction with an excess of aqueous solution of a substance which is strongly absorbed upon the fuller's earth. The solution is preferably used in an acid condition to avoid destruction of the vitamin by heat or alkalinity and is best used at a temperature of 80°C to 100°C to promote complete extraction. The fuller's earth is preferably stirred for several minutes with the hot solution which is then removed while hot by decantation, centrifugation or filtration. The extraction is preferably repeated several times and the extracts combined for further use."

Four

The Strange and Sensational Patent

Astroturf being rolled out in 1966 at the Astrodome, the first major sports venue to use the material

Astroturf

Patent Name: Monofilament Ribbon Pile Product
Patent Number: 3,332,828
Patent Date: July 25, 1967
Inventors: James M. Faria, of Decatur, Alabama, and Robert T. Wright, of Pensacola, Florida

What It Does

Astroturf simulates the look and feel of grass but with added durability, for use both indoors and outdoors for a variety of recreational and sporting activities.

Background

Walt Whitman wrote, "I believe a leaf of grass is no less than the journey-work of the stars." Certainly he could not have anticipated the invention of a factory-made synthetic grasslike turf, which was created after observers in the 1950s noted that city dwellers were less active and less fit than their country cousins. In the early 1960s, Monsanto Industries funded research to test such variables as foot traction, cushioning, and wear resistance in the development of a synthetic fiber that would serve as a viable substitute for grass playing fields. Until then, artificial turf material was developed strictly for decorative purposes. Meanwhile, Judge Roy Hofheinz wanted to build the world's first indoor sports arena. In 1966, the synthetic material, then called Chemgrass, was used to green the playing field of the Astrodome in Houston, Texas. A year later, the invention was patented and the name Astroturf became synonymous with indoor "grass."

How It Works

1. Synthetic fibers are woven on a Wilton cut-pile loom to form a woven backing with a cut-pile face extending from one side.

A recent study of professional soccer players showed that they suffered an equal number of injuries whether they played on natural grass or artificial turf, which has improved since it was originally introduced.

A new generation of Astroturf known as US Speed S5_M was installed in four NFL stadiums in 2016, including the New York Giants and New York Jets, Minnesota Vikings, Cincinnati Bengals, and New Orleans Saints.

2. A suitable latex formulation on the surface of the backing renders the structure dimensionally stable.

3. A polymeric elastomer is applied to the later backing to provide stable matting.

July 25, 1967   J. M. FARIA ET AL   3,332,828

MONOFILAMENT RIBBON PILE PRODUCT

Filed Dec. 28, 1965

FIG. 1.

FIG. 2.

FIG. 3.   FIG. 4.

FIG. 5.

INVENTORS
JAMES M. FARIA
ROBERT T. WRIGHT

BY *Roy H. Massengill*
ATTORNEY

In the Inventors' Words

"The prior art reveals that attempts to make artificial grasses have been made during the past several years. In most instances, the inventive concepts have been concerned primarily with providing a decorative artificial grass and have not attempted to provide an artificial grass which will withstand permanent outdoor installation and the abusive wear caused by spiked or cleated shoes. There are no known simulated turfs which have performance characteristics other than decorative that compare even closely with natural turf."

Bubble Wrap

Patent Name: Method for Making Laminated Cushioning Material
Patent Number: 3,142,599
Patent Date: July 28, 1964
Inventor: Marc A. Chavannes, of Brooklyn, New York, assignor to Sealed Air Corporation, of Hawthorne, New Jersey

What It Does

Bubble Wrap is a popular cushioning product used for packing and mailing fragile material throughout the world. (It is also fun to pop.)

Background

Like so many other invention tales, this one opens with a pair of buddies tinkering around in a garage. In 1957 in Hawthorne, New Jersey, with geek dreams in their hearts and odd materials at their disposal, Marc Chavannes and Al Fielding were trying to make some kind of plastic wallpaper. Instead, they came up with a laminated material that had a lot of bubbles in it. They soon realized its potential as a cushioning product for packaging, and they were off. In 1960, they established Sealed Air Incorporated, which continues to make Bubble Wrap in addition to an impressive line of other products used for packaging. The original material revolutionized the shipping industry—and provided hours of irritating popping fun to recipients along with their packages.

How It Works

Conceived in a garage in New Jersey, the Sealed Air Corporation now includes thirty worldwide labs researching and producing a variety of materials completely unrelated to the original inventors' wallpaper aspirations.

1. Polyethylene resin is extruded through a cylinder with a screw inside that runs its length. As the screw turns, the resin is heated and melted.
2. The resulting liquid is formed into clear plastic film and placed in two stacks.
3. Depicted in patent illustration 9, one layer is wrapped around a drum (10) with circular pockets or other shapes punched into it.
4. Within the drum, vacuum suction draws the film into the holes that form the bubbles.

5. The second layer of film is then laminated, sealing the air inside the holes.
6. Patent illustration 6 depicts the layer (11) with embossed hexagonal pockets (24), which is hermetically sealed to its bottom film layer (16).

July 28, 1964   M. A. CHAVANNES   3,142,599

METHOD FOR MAKING LAMINATED CUSHIONING MATERIAL

Filed Nov. 27, 1959   3 Sheets—Sheet 2

In the Inventor's Words "Another aspect of the invention resides in an improved method and apparatus for the fabrication of an improved laminated cushioning material wherein at least one lamina or layer is formed to provide a plurality of discrete elements and then a second layer is hermetically sealed to the formed layer to seal the elements and thereby provide sealed pockets in which air or other fluid is entrapped. Since the laminated layers of material have flexibility and elasticity, the resultant laminated material affords a high degree of cushioning as well as shock absorbing action. In the fabrication of such cushioning material, the pneumatic cushioning may be supplemented by an improved mechanical cushioning effected by the design and configuration of the formed elements."

Chicken Goggles

Patent Name: Eye-Protector for Chickens
Patent Number: 730,918
Patent Date: June 16, 1903
Inventor: Andrew Jackson, Jr., of Munich, Tennessee

What It Does   When chickens are cooped, they tend to attack one another. Eye protection helps to prevent them from being blinded by the pecking. (The goggles are not intended as a vision aid for fowl.)

Background   This seemingly outlandish invention actually had a very practical purpose: to protect chickens from one another. When chickens are cooped together and one of them sustains some sort of blood-shedding injury, its fellow chickens tend to go into a frenzy. If one chicken has blood on its plumage, its neighbor will attack, spattering more blood and igniting a chain reaction of chicken terror. The birds use their pointed beaks as weapons, oftentimes blinding, maiming, and killing one another. Many small chicken farm operations at the turn of the twentieth century had their own methods of safeguarding their fowl, and eye protection was not uncommon.

In the 1930s, a company began marketing chicken glasses with rose-colored lenses. The reddish tint camouflaged the sight of the blood that tended to drive the chickens nuts. Today, most large chicken and poultry operations keep their birds separated. At free-range chicken farms, the pointed beaks of chicks are snipped at an early age to prevent carnage and mutilation of the poultry. But you can't help loving this chicken eyewear, which lent a certain professorial aura to the bird, belying its innate bloodlust and brutality.

How It Works   Resembling a miniature pair of Lennon specs, this invention consists of two circular frames connected by a loop that fits over the chicken's beak. A U-shaped strap attached to the loop where it meets the circular frames fastens around the chicken's head.

No. 730,918.
PATENTED JUNE 16, 1903.
A. JACKSON, Jr.
EYE PROTECTOR FOR CHICKENS.
APPLICATION FILED DEC. 10, 1902.
NO MODEL.

Fig. 1.

Fig. 2.

Inventor:
Andrew Jackson Jr.
By. Chandler & Chandler
Attorneys

Witnesses
O. McIngIngram
Harry Ellis Chandler

In the Inventor's Words

"This invention relates to eye-protectors, and more particularly to eye-protectors designed for fowls, so that they may be protected from other fowls that might attempt to peck them, a further object of the invention being to provide a construction which may be easily and quickly applied and which will not interfere with the sight of the fowl."

Escapable Coffin

Patent Name: Improved Burial Case
Patent Number: 81,437
Patent Date: August 25, 1868
Inventor: Franz Vester, of Newark, New Jersey

"The boundaries which divide Life from Death are at best shadowy and vague. Who shall say where the one ends, and where the other begins? We know that there are diseases in which occur total cessations of all the apparent functions of vitality, and yet in which these cessations are merely suspensions, properly so called."
—Edgar Allan Poe, "The Premature Burial"

What It Does   Provides a means of escape for anyone finding himself in the unfortunate circumstance of having been buried alive.

Background   Being buried alive certainly sounds like the macabre stuff of Poe, but such misfortunes were probably not completely uncommon prior to the twentieth century. Because medical means for determining death were less sophisticated than they are today, coma victims, for example, ran the risk of being sent to their graves prematurely. Franz Vester's invention was not the first in a series of "safety coffins" with escape hatches or signaling means, nor was it the last. But, certainly it was among the most elaborate.

Death Be Not Proud. Or Cheap. The average casket sold in the United States now costs slightly more than $2,000; some sell for as much as $10,000, according to the Federal Trade Commission.

"What would Mozart be doing if he were alive today? Scratching at the door to his coffin!" Not so, if he had been buried in Franz Vester's "improved burial case," patented in 1868.

F. Vester.
Coffin.
Nº 81,437.   Patented Aug. 25, 1868.

How It Works

1. The square tube (C) extends from above the surface of the grave to a base (D) on the lid (B) of the coffin; and includes air-inlet openings (F) that connect into the coffin.

2. At the base on the lid is a sliding glass door (L), held in place by a spring bar (E), which allows investigators, or the morbidly curious, a peek at the face inside the casket from above ground.

3. Inside the tube is a ladder (H), and on top of the tube is a bell (I).

4. Attached to the bell is a cord (K), which extends the length of the tube and into the coffin, where its end is laid in the buried person's hand.

5. Inside the coffin lid, the sliding glass door is ready to be closed by a spring (M) activated once the tube is withdrawn—when the deceased is determined to be, with unwavering certainty, deceased and ready, at last, to rest in peace.

6. When the glass door shuts, air is closed off from the coffin, and the tube can be reused.

7. Ashes to ashes, dust to dust; if death be uncertain, this coffin is a must.

In the Inventor's Words

"The nature of this invention consists in placing on the lid of the coffin, and directly over the face of the body laid therein, a square tube which extends from the coffin up through and over the surface of the grave, said tube containing a ladder and a cord, said cord being placed in the hand of the person laid in the coffin, and the other end of said cord being attached to a bell on the top of the square tube, so that should a person be interred ere life is extinct, he can on recovery to consciousness, ascend from the grave and the coffin by the ladder; or, if not able to ascend by said ladder, ring the bell, thereby giving an alarm, and thus save himself from premature burial and death; and if, on inspection, life is extinct, the tube is withdrawn, the sliding door closed, and the tube used for a similar purpose..."

Genetic Engineering

Patent Name: Process for producing biologically functional molecular chimeras
Patent Number: 4,237,224
Patent Date: December 2, 1980
Inventors: Stanley Cohen, of Portola Valley, California, and Herbert Boyer, of Mill Valley, California

What It Does   Manipulates DNA structure by incorporating replicable genes into cells.

Background

The first cloned human embryos were created in 2001 by a team of Massachusetts-based scientists who hope to eventually produce cloned stem cells to cure disease.

Founded in 1997, a fringe group calling itself Clonaid claims to have successfully cloned several human babies. The company was established by supporters of the Raëlian Movement, described on the company website as "the world's largest UFO-related organization." The group takes its name from its spiritual leader, "His Holiness Raël," who promotes cloning as the way to eternal life. Their claims have yet to be verified.

The science of genetics is the science of heredity and of the mechanics by which characteristics are passed from one generation to the next, whether from virus to virus, fruit fly to fruit fly, or obnoxious parent to obnoxious child. The seed for the concepts of genetics was planted by the Augustinian monk Gregor Mendel in 1861, when he showed that hereditary traits in pea plants were transmitted from one generation to the next in predictable patterns.

Much later it was shown that Mendel's laws applied to all species, including humans. However, in many cases the inheritance of genes turned out to be much more complicated than Mendel had theorized. The inheritance of complex traits was shown to be due to the interaction of several Mendelian factors or genes.

In the 1940s, experiments with bacteria established that deoxyribonucleic acid, or DNA, is the chemical basis of the gene. Enter James Watson and Francis Crick, who are credited with discovering the structural basis of genetic information in DNA and opening the floodgates in the field of genetics and the many branches of the science that have flourished from it.

In 1953, James Watson and Francis Crick described the structural basis of genetic information in DNA as a double helix. This structure resembles a ladder twisted in a clockwise spiral. The order of nucleotides (the crossbars of the double helix) along the DNA molecule was found to specify the linear arrangement of amino acids in proteins. Thus, the genetic code governs the translation from nucleotide order to amino acids to proteins to the very structure of living organisms as far apart as viruses and humans.

The genetic code functions by the rule that triplets of DNA nucleotides each specify a single amino acid. Restriction enzymes also each specify a single amino acid. Restriction enzymes cut DNA at specific sites and allow the in vitro recombination of DNA within and between organisms and between species. These and other discoveries led to the development of molecular biology, and eventually to genetic engineering.

By early 1973, Herbert Boyer and Stanley Cohen added replicable genes to a simple cell, giving birth to recombinant DNA. Their findings have not only been of great fundamental significance but are increasingly applied in many branches of genetic biology, including agriculture, animal breeding (hello, Dolly!), and medicine. Genetic engineering, or gene manipulation, is now a widely debated experimental technology, developed to alter the genome of a living cell for medical or industrial use. Indeed, the method described in this patent has opened the doors to a strange new world.

Dolly, the first genetically engineered sheep

A Finn Dorset sheep was born in July 1996. She was named Dolly after country singer Dolly Parton, and she became just as famous as her namesake. She was cloned from the breast cell of an adult ewe. She matured and mated naturally with a mountain ram. Dolly gave birth to a healthy lamb and to the ongoing debate on the ethics of human cloning.

How It Works

1. Plasmids or virus DNA are cleaved to provide linear DNA having ligatable termini.

2. A gene possessing complementary termini is inserted, providing a biologically functional replicon with a desired phenotypical property.

3. The replicon is then inserted into a microorganism cell by transformation.

4. Isolation of the transformants provides cells for replication and expression of the DNA molecules present in the modified plasmid.

In the Inventors' Words

"The ability of genes derived from totally different biological classes to replicate and be expressed in a particular microorganism permits the attainment of interspecies genetic recombination. Thus, it becomes practical to introduce into a particular microorganism, genes specifying such metabolic or synthetic functions as nitrogen fixation, photosynthesis, antibiotic production, hormone synthesis, protein synthesis, e.g. enzymes or antibodies, or the like—functions which are indigenous to other classes of organisms—by linking the foreign genes to a particular plasmid or viral replicon."

Goat Traction Footwear

Patent Name: Footwear with Mountain Goat Traction Elements
Patent Number: 6,226,896
Patent Date: May 8, 2001
Inventor: Michael Ray Friton, of Portland, Oregon; assigned by Nike, Inc.

What It Does

The shoe invention described in this patent takes as its model the impressively sure-footed mountain goat.

Background

One of the most adept mountain climbers in North America, the desert bighorn sheep is now trying to climb its way off of the endangered species list.

A former University of Oregon college runner named Phil Knight went into the sneaker business in 1962. He adapted a waffle-shaped sole, a cushioned midsole, a wedged heel, and nylon tops, outstripping his predecessors in the sport shoe business with the lightness and flexibility of his product. A decade later, he was marketing shoes named after the Greek goddess of victory, Nike.

Because of the growing interest in trekking the outdoors, an array of hiking shoes and bulky boots began lining the shelves of shoe stores worldwide in the eighties. Lightweight trail shoes and durable sandals soon followed, including the Nike Mada and Terra trail shoes and the Nike Terra and Deschutz.

How It Works

Just Do It, Kid. Nike received a great deal of public criticism for exploiting child labor in poor countries, such as Pakistan, Cambodia, and Vietnam. The company has admitted to some of the allegations, raised wages and increased factory audits, and published public reports on its improvements.

1. The sole has a ground-engaging surface including soft, compliant traction elements and one or more relatively hard lugs.
2. The lugs are stiffer in compression than the traction elements and run adjacent to them.
3. The traction elements extend downwardly below the lugs such that, in use, a bottom surface of the traction elements will make initial ground contact and partially compress.
4. The compression cushions the impact of the ground engagement and increases the ground contact. The bottom surface of the lugs touches the ground.

5. The lugs limit compression of the traction elements and serve as a relatively rigid catch for irregular and soft ground surfaces.
6. Cushioning is attached to the upper.
7. The sole has a ground-engaging surface including an outer perimeter border and an interior.
8. The interior region comprises a group of relatively soft traction elements.
9. The border region comprises a pair of relatively hard lugs, stiffer and adjacent to the group of traction elements.

FIG. 3

In the Inventor's Words "The effectiveness of a mountain goat's hoof in providing that animal with sure footing on diverse and extreme ground conditions has been recognized. As described in the book entitled *Beast the Color of Winter, the Mountain Goat Observed*, by Douglas H. Chadwick, Sierra Club pub. (1983), '[t]he sides of a mountain goat's toes consist of the same hard keratin found on the hoof of a horse or deer. Each of the two wrap-around toenails can be used to catch and hold to a crack or tiny knob of rock…The mountain goat is shod with a special traction pad which protrudes slightly past the nail. This pad has a rough textured surface that provides a considerable amount of extra friction on smooth rock and ice. Yet it is pliant enough for any irregularities in a stone substrate to become impressed in it and thereby add to the skid proofing effect.'"

Lava Lamp

Patent Name: Display Device
Patent Number: 3,570,156
Patent Date: March 16, 1971
Inventor: Edward C. Walker of Hampshire, United Kingdom

What It Does

The lava lamp is an ornamental lamp consisting of a transparent glass vessel and a visually pleasing lavalike substance that continually bubbles to the surface and then flows back down when the lamp is on.

Background

Edward C. Walker created the lava lamp while trying to redesign a wax-based egg timer he had seen in a pub.

Edward Craven Walker once described the spectacle of his invention as a sexy thing that "starts from nothing, grows possibly a little bit feminine, then a little bit masculine, then breaks up and has children." A World War II veteran, Walker adopted the lingo and the lifestyle of the flower children. Part Thomas Edison, part Austin Powers, he was a nudist in the psychedelic days of the UK—and possessed some pretty savvy marketing skills to boot. "If you buy my lamp, you won't need to buy drugs," he was known to say.

In the mid-'60s, Walker hit on the idea of using a lamp concealed within a base to display a groovy glass vessel that contained a lavalike oil and wax substance swimming in a colored liquid. As the lamp heated the substance in the globe, the lavalike substance would change form and shape. At first, shopkeepers resisted the strange lamp. But the lava lamp was soon illuminating nightclubs and flats across London.

Eventually, a pair of enterprising Americans bought rights to manufacture the lava lamp in the U.S. Their company slapped on the Lava Lite® trademark, and the lamps have been selling continuously ever since.

How It Works

1. Dyed water and another liquid of mineral oil, paraffin wax, paraffin, and carbon tetrachloride is sealed into a glass container.
2. The container rests in a hollow base, which contains an electric lightbulb.
3. When the lamp is off and cold, the second liquid congeals in a hard lump at the bottom of the glass.
4. When the lamp is on, the substance expands, loses density, and rises.
5. As it rises, it moves farther from the heat source and begins to resolidify and sink.
6. This heating and cooling process is constantly reshaping the substance swimming in the dyed liquid as it is illuminated by the lamp.

In the Inventor's Words

"I think it will always be popular," Walker told the Associated Press before his death in 2000. "It's like the cycle of life. It grows, breaks up, falls down, and then starts all over again."

Maglev

Patent Name: Electromagnetic Inductive Suspension and Stabilization System for a Ground Vehicle
Patent Number: US 3470828 A
Patent Date: October 7, 1969
Inventors: James R. Powell, Jr., and Gordon T. Danby, of Upton, New York

What It Does   This transportation technology uses magnetic fields to levitate and propel a vehicle, most often a train.

Background   Much of the pioneering work on magnetic levitation, or maglev, was done by Eric Laithwaite, who built the first working full-size linear induction motor. Laithwaite realized that the linear motor, which does not require physical contact between the vehicle and the track, could be used to develop a system of transportation based on magnetic fields. He went on to experiment with single linear motors that could use an arrangement of magnets to produce both lift and forward thrust.

Many organizations took notice of the technology, and James Powell and Gordon Danby of Brookhaven National Laboratory received the first patent for a maglev train, in 1967. The patent outlines "a high-speed train utilizing superconducting magnets...This system generates a suspension force, for floating the train above the ground." By lifting the vehicle off the track with electromagnets, the force of friction can be nearly eliminated, and a small amount of energy is sufficient to accelerate the vehicle to high speeds. The original patent for a maglev train suggests using a "propeller, jet, [or] rocket" to propel the vehicle. However, it was not until Laithwaite's work on linear induction motors was married to Powell and Danby's vehicle design that the first practical maglev trains were born.

The first magnetic levitation passenger vehicle, simply called Maglev, was a low-speed shuttle that opened in 1984 in the United Kingdom. It ran between Birmingham International Airport and the nearby railway station. The track was only 2,000 feet long, but the train cars were successfully levitated about 15 millimeters (0.6 inches) by powered electromagnets. Linear motors were used to propel the vehicle up to 26 mph.

Although the UK's first maglev shuttle was shut down in 1995 due to reliability issues, maglev trains were constructed in other countries in the 1980s and '90s. In Germany, for example, a number of experimental maglev trains were built and tested, resulting in the high-speed Transrapid train. A German-developed Transrapid now serves as the Shanghai Maglev Train, which is the fastest passenger train in service, with a top operating speed of about 270 mph. Japan and South Korea have also constructed high-speed maglev trains for commercial transportation, sometimes called "bullet trains." On a test run, Japan's L0 Series maglev train hit 375 mph on the Shinkansen high-speed railway in April 2015, setting the world record for train speed.

In the future, a number of companies, such as Hyperloop One and Hyperloop Transportation Technologies, hope to build a commercial transportation system that uses magnetic levitation to accelerate a passenger pod inside a vacuum-sealed tube. In the vacuum chamber, with virtually no air resistance, the pods could theoretically accelerate to speeds over 750 mph. A Hyperloop system was first imagined and described by Elon Musk in 2013.

How It Works

1. Maglev trains use superconducting magnets, which are electromagnets that have been cooled to below –450° Fahrenheit to increase the power of the magnetic field by about a factor of ten.
2. Superconducting magnets are installed in the train cars themselves, as well as in the tracks.
3. Electric currents supplied to the magnets generate the required electromagnetic fields to levitate and propel the trains.
4. The trains are usually levitated above a concrete guideway.
5. Metallic loops set into the concrete, made from conductive materials like aluminum, interact with the magnetic field formed by the train and generate additional magnetic fields.
6. Three primary magnetic fields are generated to levitate the train about 5 inches above the concrete guideway, keep it centered on the track horizontally, and propel the train.
7. The magnetic field used for propulsion undergoes alternating currents to both push the train car from behind and pull it from the front.
8. Even at speeds in excess of 250 mph, the ride is very smooth, because the only source of friction is the air.

9. The entire system runs according to computer algorithms, with no one "driving" the train.

10. Maglev trains are designed so the farther the train strays from the optimal position, the stronger the magnetic forces pushing it back are, just like when you hold the same pole of two magnets close together. This significantly reduces the chances of these trains derailing compared to conventional trains.

In the Inventor's Words "In 1959, I was on my way to see a girlfriend up in Boston, Massachusetts, and I got stuck in a 5 hour traffic jam on the Throgs Neck Bridge. And I was just sitting there in the car and I thought, 'There must be something, a better way to get to Boston.' And I thought, 'Well, there's trains and so forth.' And then I thought, 'Gee, what if you used superconducting magnets to levitate vehicles and have them travel along a track?'" —James Powell

Times Square, New York City, 1934

Neon

Patent Name: System of Illuminating by Luminescent Tubes
Patent Number: 1,125,476
Patent Date: January 19, 1915
Inventor: Georges Claude, of Paris, France

What It Does

Neon gas fills tubes to provide a cool, long-lasting, oddly brilliant nonincandescent lighting tailor-made for signs that must be seen from a long distance.

Background

Neon is a rare, inert gas found in very small quantities in the air. In the late eighteenth century, scientists used glass tubes to observe the results of experiments that involved high-voltage charges. After the first five noble gases—helium, neon, krypton, argon, and xenon—were discovered in the 1890s, Georges Claude observed that electrical discharges passing through some of these gases could produce light without incandescence. Neon was the gas that produced the brightest light.

In 1912, the world's first neon sign illuminated the words "Le Palace Coiffeur" above a Paris barbershop. In that same year, Cinzano Italian vermouth became the first product advertised by neon lighting.

Viva Las Vegas. Las Vegas is the ultimate neon city, luring tourists and their dollars from all over the world to see what has been called the first truly postmodern city in the world.

In 1910, Claude introduced the first neon light display in Paris. In 1923, the French scientist-turned-entrepreneur sold two neon signs to a Los Angeles car dealership, and within a year, he owned a chain of "Claude Neon" stores throughout the United States. As the neon lighting industry grew more sophisticated, neon signs became a regular feature of the urban landscape, adorning movie theaters, car dealerships, bowling alleys, restaurants, and bars.

This patent was awarded to Claude five years after his first French patent for neon lighting and represents a refinement in his process. He included methods in this patent to obtain a more brilliant neon light, and a way to keep gaseous deposits at the electrodes from altering the color of the tube.

How It Works

1. A vacuum pump (c) removes air from a glass tube bent into the desired shape.
2. Another tube (d) inserts the neon gas.
3. A carbon receptacle (b) is immersed in liquid air (e) to facilitate removing excess foreign gases and give the tube its fullest possible brilliancy.

4. The lamp is then separated from the carbon receptacle.

5. The tube (a) contains electrodes (g) at either side to receive electrical current and illuminate the gas within the tube.

6. When a neon lamp is plugged into a source of electricity and turned on, the gas becomes illuminated.

In the Inventor's Words

"What I claim and desire to secure by Letters Patent of the United States is… [a] luminescent tube containing previously purified neon and provided with internal electrodes for illuminating said gas, said electrodes being deprived of occluded gases and having an area exceeding 1.5 square decimeters per ampere, to decrease the vaporization of the electrodes and prevent the consequent formation upon the walls of the tube, in proximity to said electrodes, of deposits containing said gas, whereby the luminosity of the tube is maintained constant for a very considerable period of time without a fresh introduction of gas."

Ouija Board

Patent Name: Toy or Game
Patent Number: 446,054
Patent Date: November 10, 1891
Inventor: Elijah J. Bond, of Baltimore, Maryland, assignor to Charles W. Kennard and William H. A. Maupin, both also of Baltimore, Maryland

What It Does

The Ouija board is a novelty used for conducting séances and communicating with the dead.

Before there were television psychics, there was the Ouija.

Background

Is there really life after death? Yes, at least for the Ouija. Today, the Parker Brothers Company sells Ouija boards that are reproductions of the spirit-summoning board sold more than one hundred years ago. They hold all of the patents and trademarks to the board and they sell lots of them. How many? You'll have to ask the Ouija.

To Infinity and Beyond. You can talk to spirits in cyberspace with an online Ouija board at many websites.

According to the Museum of Talking Boards, the year 1848—the year before Edgar Allan Poe died—was pivotal in the rise of modern spiritualism. That year, in a New York cabin, two sisters claimed to have contacted the spirit of a dead person and started a fervor for the occult in the United States and Europe. Spiritual mediums, crystal-ball gazers, palm readers, and a motley crew of others rose like zombies from the grave. One popular method for conversing with "the other side" was through table turning. A specialist would conduct a séance in which attendees placed their fingers on a table edge; upon contact, the table might tip in one direction or the other as a mounted dial-like mechanism pointed to letters and spelled out messages. The stage was set for the introduction of the Ouija board.

Elijah Bond was acquainted with a shrewd group of men who were no doubt aware of the public interest in the occult; they lived in the very city where Poe was buried. Certainly, they concluded, there was more of a market for the supernatural than just books and carnival hoodoo gimmickry. The Kennard Novelty Company was established a year before this patent was issued and became the first company to commercially manufacture the Ouija board. Its precise cultural origin remains a subject of debate, but the board's promise of a means to communicate with the underworld was very attractive to many, in spite of the fact that the patent itself clearly character-izes the item as a "Toy or Game."

After company power struggles, ownership of the Ouija board product changed hands several times and legal battles ensued. The owners apparently lost a battle with the Internal Revenue Service when they tried to claim that their product was a spiritual tool and, therefore, its sales should not be taxed. In the late 1960s, a new age of spiritualism sparked new interest in, and even fear of, the talking board. It was then, too, that ownership of the Ouija board went to Parker Brothers, which was bought by Hasbro, based in Salem, Massachusetts, where a number of burned witches probably still have a lot to get off of their charred chests.

How It Works　A flat-surfaced board approximately 15 by 22 inches in size features the letters of the alphabet in two arching rows, the symbol "&" below the letters, and below this, the digits 1, 2, 3, 4, 5, 6, 7, 8, 9, and 0 straight across. Also printed on the board are the words "Yes," "No," and "Goodbye" with accompanying illustrations. A stout table-like structure (D) features an elongated section (E). The little table is placed on the board. The hand of the player is placed lightly on the table. Questions are posed. The leg beneath the elongated portion of the table will point to letters that may be interpreted as answers.

In the Inventor's Words　"My invention relates to improvements in toys or games, which I designate as an 'Ouija or Egyptian luck-board;' and the objects of the invention are to produce a toy or game by which two or more persons can amuse themselves by asking questions of any kind and having them answered by the touch of the hand, so that the answers are designated as letters on a board."

Pigeon Starter

Patent Name: Improvement in Pigeon Starters
Patent Number: 159,846
Patent Date: February 16, 1875
Inventor: Henry A. Rosenthal, of Brooklyn, New York

What It Does

Startles pigeons from traps for an improved recreational shooting experience.

Background

There's a story of an English sportsman who, during an especially frustrating day afield, decided to tie one of his pigeons to a stool, thereby making it a live decoy to attract other pigeons. Hence, the term "stool pigeon" refers to a police informer, a person who can bring trouble near.

In the nineteenth century, sportsmen enjoyed target-shooting with live pigeons, which were often kept in traps. When their owners wished to take target practice, they would bring the cage to an open field and unlatch the door. The problem was that pigeons were often reluctant to fly out of the trap; either the birds were not too bright, or on the contrary, they possessed an intuition as to what this new freedom really might mean. Enter the pigeon starter—a likeness of a natural predator meant to startle the pigeon out of the trap and into flight.

The pigeon starter is today no more than an historic curiosity. With automated clay trap dispensers, trapshooting remains a popular sport without contributing to any loss of avian life.

How It Works

1. A figure of a cat or other animal (A) includes four rigid legs that are pivoted at their upper ends.

2. The feet of the animal are attached securely to a baseboard (B) designed to be staked to the ground.

3. Between the fore and rear legs, a coiled spring (C) is secured to the baseboard and is attached to the lower end of a lever (D).

4. The upper end of the lever is secured to the underside of the animal's body.

5. A catch (E) is attached to the base and held forward by a spring (F) also attached to the baseboard in order to hold the lever when the animal is pressed down.

6. A cord (G) is attached to the end of the spring catch (E, F), which, when pulled, activates the mechanism and abruptly launches the animal likeness into an erect position.

According to the Smithsonian Institution, the passenger pigeon once constituted 25 to 40 percent of the total bird population of the United States. At the time Europeans first came to America, there were an estimated three to five billion passenger pigeons. Overhunting led to the decline of the species, and at the time of this patent, it was in serious peril. On September 1, 1914, the species became officially extinct when the last known passenger pigeon died at the Cincinnati Zoo.

"As traps have heretofore been used, it frequently happens that the pigeons will not leave the trap when it is sprung, and have to be frightened out by shouting or throwing stones, &c., which tends to make the sportsman nervous, and frequently causes him to lose his shot."

"This pigeon starter is formed by the combination of the body, the rigid pivoted legs, the spring, the lever, the spring catch, and the trip-cord with each other and a base-board; and in the employment, as a pigeon-starter, of a figure made to represent the body of an animal, and provided with suitable mechanism to enable it to be set in a crouching position, and to be sprung into an erect position…"

Leon Theremin with his invention, 1927

Theremin

Patent Name: Method of and Apparatus for the Generation of Sound
Patent Number: 1,661,058
Patent Date: February 28, 1929
Inventor: Leon Theremin, of Leningrad, Russia

What It Does   This electronic music-making device emits haunting tones when "played" by moving one's hands in the vicinity of its active parts, two radio-frequency oscillators.

Background   Leon Theremin was born Lev Terman in Russia in 1896. One day, he was putting together parts for a radio when he began hearing noises he knew were not part of any radio broadcast. He soon realized he could manipulate and control the frequencies of the sounds by waving his hands near the device. From these initial observations of the interruption of sound waves, Terman

A 1993 documentary film, *Theremin: An Electronic Odyssey*, offers great insight into the inventor's strange and fascinating life. In it, Robert Moog, who went on to create the Moog synthesizer and other innovations in electronic sound, describes his own early interest in the theremin, dozens of which he made as a kid.

In addition to the Beach Boys' use of the theremin in "Good Vibrations," Led Zeppelin, a hugely popular band in the '70s, was known to use a theremin from time to time. More recently, the Jon Spencer Blues Explosion featured the instrument in the 1994 album *Orange*, and even featured a rendering of the instrument on its album cover.

invented the world's first electronic musical instrument, aptly named the "theremin," in 1919. The invention was first publicly demonstrated at the Moscow Industrial Fair in 1920. Lenin himself commissioned six hundred of the instruments to be built and toured around the Soviet Union.

Theremin (as he renamed himself) emigrated to the United States and soon succeeded in showcasing his musical ingenuity. Audiences everywhere were amazed as he played the theremin. Here was a musical instrument that was played without physical contact, by manipulating the energy currents in the air around it with the movement of the hands: One hand controlled volume; one controlled pitch. The result was an eerie violin-like sound with a haunting beauty and magical quality. The invention was a marvel of both science and art, and the earliest example of electronic music. Today the sounds of the theremin are recognizable to many as the high-pitched quavering notes used to score the sci-fi flicks of the fifties, and used by the rock group the Beach Boys, most notably in their song "Good Vibrations." But the sound of the theremin is no more bizarre than the history of the man who created it.

The inventor settled in New York City, living large with lots of friends and admirers. He went on to develop a variation of the instrument that could be played by the movement of the human body, so that dance motions could be translated into music. There was even a troop of "theremin dancers," as they were called. In 1938, he suddenly and inexplicably disappeared, only to reappear a half century later.

Alexandra Stepanoff, Leon Theremin's first student in the United States

Feb. 28, 1928.

L. S. THÉREMIN

1,661,058

METHOD OF AND APPARATUS FOR THE GENERATION OF SOUNDS

Filed Dec. 5, 1925

6 Sheets—Sheet 1

Fig.1.

Fig.1a.

Fig.2.

Fig.3.

Inventor:
Leo Sergejewitsch Theremin
By Dowell and Dowell
his Attorneys

Theremin had been kidnapped from his New York apartment and taken to the Soviet Union, where he was made to work on a variety of secret government projects, including the development of the "bug" for the KGB. Eventually, in his old age, he reunited with some of his friends from the United States and enjoyed the recognition of his contributions to electronic music. He died in Moscow in 1993, at the age of 97.

How It Works   The theremin is by no means an easy instrument to play; there are no specific reference points for the musician to read at all. Sound is produced via radio-frequency oscillators. One oscillator is set at a fixed frequency; another, at a variable frequency. One hand moves around the metal loop controlling volume. The other hand controls pitch by adjusting its proximity to the antennae.

In the Inventor's Words "An instrument embodying the invention comprises a sound reproducer, such as a telephone receiver or loud speaker, connected to an oscillating system adapted to be controlled or affected by an object or objects, such as the hands or fingers of an operator held in relative position in proximity to an element of the system."

"Towering Inferno" Fire Blanket

Patent Name: Apparatus for Extinguishing Fires in High Rise Block Buildings or Uniform Transverse Cross-Section Plan
Patent Number: Great Britain 1,453,920
Patent Date: October 27, 1976
Inventor: Arthur Paul Pedrick, of Selsey, Sussex, United Kingdom

What It Does This innovation in fire control was intended to provide an alternative method to extinguish fires in tall buildings, thus minimizing the casualties and injuries in a tragic "towering inferno" situation.

Background Disaster movies were big in the early 1970s, and the 1974 release of *The Towering Inferno*, starring Steve McQueen and Paul Newman, among other luminaries, was the king of the genre. The trend may have triggered the imaginative spark of inventor Arthur Paul Pedrick, who set about devising a disaster-aversion scheme. Listing himself as the subject of a "One-Man Think-Tank" in this British patent application, Arthur Pedrick describes one of his many intriguing inventions: A giant fire-resistant blanket that rolls down from the top of a tall building to smother a fiery conflagration.

How It Works 1. Spindles are secured to the rooftop of a building.

2. Around them are spun rolls of fire-resistant material with weights at their ends.

3. In the case of a fire, the curtains may be released manually or by a thermoresponsive switch.

4. Guide ducts secured along the building may be used to help align the weights and draw the curtain down steadily.

5. To avoid the potential of suffocating people inside the building, the blanket includes strategically placed holes that align with windows in designated "fire emergency" rooms.

Some of Pedrick's many patent applications:

- Earth Orbital Bombs as Nuclear Deterrents
- Reducing the Tendency of a Golf Ball to Slice or Hook by Electro-static Forces
- Automobiles Driven from the Back Seat
- Aerial Ships Supported by Vacuum Balls or Other Forms of Evacuated Vessels
- Speed of Light Regulated Clock
- Photon Push-Pull Radiation Detector for Use in Chromatically Selective Cat Flap Control
- 1000 Megaton Earth-Orbital Peace-Keeping Bomb

In the Inventor's Words "It is generally impracticable to construct fire escape ladders of over 200 ft. in height…[W]hen fires have occurred in high rise buildings, people cut off from the lifts or fire escape ladders by the fires starting below them have been forced to jump for their lives, if they are not rescued in time by a helicopter landing on the roof."

Zeppelin

Patent Name: Navigable Balloon
Patent Number: 621,195
Patent Date: March 14, 1899
Inventor: Ferdinand Graf von Zeppelin, of Stuttgart, Germany

What It Does

The zeppelin dirigible is a lighter-than-air craft with a rigid frame designed as a vehicle for passengers or freight.

Background

The spire on top of the Empire State Building was originally planned to be used as a mooring mast for dirigibles. Gusty wind conditions at 1,350 feet rendered the plan unworkable, but it is fun to imagine looking up and seeing blimps docked over New York City.

A 1969 groundbreaking rock-and-roll debut album featured on its cover the ominous and shadowy underside of the *Hindenburg* as its back end exploded in flame. The album, like the band, was called *Led Zeppelin*.

Long before the Wright Brothers took flight in 1903, another pair of brothers achieved a feat of similar magnitude. The French Montgolfier brothers demonstrated the first flight of a model hot-air balloon in 1783. Others in France that year were quick to follow, and human flight was, in effect, achieved. Over time, different styles and models of balloon-like aircraft were conceived, including dirigibles, which were self-propelled lighter-than-air craft and featured directional control surfaces. One inventor of such airships is now permanently linked by name to his invention: Ferdinand Graf von Zeppelin.

A former Prussian military officer, Zeppelin built his first rigid dirigible—composed of a sturdy lightweight framework—in 1900. Filled with hydrogen gas, the ship was approximately 420 feet long and 40 feet in diameter. A pair of gondolas carried the passengers and crew. Rudders at the front and back provided steering, and an internal combustion engine was used to propel it.

But Zeppelin had bigger ideas: fleets of the large airships moving great quantities of freight. In 1908, he created the Zeppelin Foundation to further his ideas in aerial navigation. Two years later, one of his dirigibles was licensed to provide the first commercial passenger air service of its kind.

The largest airship made by Zeppelin's company was named the *Hindenburg*. It is said to have been designed to reflect the greatness of Nazi Germany. Stretching more than 800 feet—the length of several city blocks—the luxury passenger craft offered two-day trips from Europe to North America. The *Hindenburg* would later be compared to the *Titanic*, not only for its size but for its tragic fate. On May 6, 1937, in Lakehurst, New Jersey, the *Hindenburg*, carrying passengers from Europe, burst into flames just as it was docking, and thirty-five people died in the tragedy. The event marked the abrupt end of the golden era of the passenger airship, and has lingered in memory, as it was captured dramatically on film.

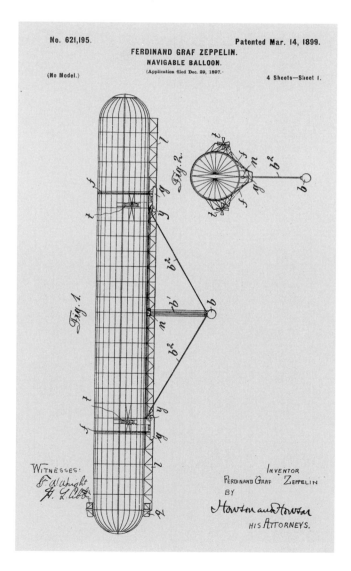

How It Works   A framework is divided into separate compartments. A main gasbag and auxiliary gasbags are contained within each compartment. A moveable carriage supports the weight with rotary drums and also attaches to the rope and the gondolas. A series of balloons coupled together are

In 1997, the Zeppelin Company built a new airship that is certified for passenger flight.

provided with rigid casings, the front casing being provided with a driving mechanism. With coverings securing the casings and the spaces between the paired balloons, the remainder of the aircraft is adapted to carry the load or freight.

1. A running weight (b) is suspended by pulley ropes and tackle (b^1, b^2) beneath the ship.

2. Two gondolas (g) are suspended beneath the ship as well.

3. The pulley ropes are accessible from the gondola to manipulate the weight to maintain the ship's balance as it rises or moves forward.

4. Also suspended beneath the ship, a gangway (l) extends almost the length of the entire ship.

5. Operators have full access to the ship by rope ladders (f) to make adjustments to air.

6. A pair of rudders (q) at one end of the ship is used in steering.

In the Inventor's Words

"This invention relates to a navigable balloon which is characterized essentially in that it is provided with a number of motors arranged separately from each other. In this manner it is possible to give the balloon or buoyant part of the apparatus, which receives the gas and is preferably cylindrical with rounded ends, a smaller diameter in proportion to the driving power developed by the motors and to correspondingly reduce the air resistance. A navigable balloon or air craft of this kind can be combined with several other balloons or air crafts in such a manner that the foremost craft contains the driving gear, while the others serve for the reception of the goods or loads to be carried."

Five

The Patent That Eases

Fig. 1.

Automated Teller Machine

Patent Name: Credit Card Currency Dispenser
Patent Number: 3,761,682
Patent Date: September 25, 1973
Inventors: Thomas R. Barnes, of Dallas, Texas; George R. Chastain, of Irving, Texas; and Don C. Wetzel, of Dallas, Texas

What It Does

Dispenses cash, receives bank deposits, and relays account information to bank customers automatically, by means of public terminals activated by personal magnetic cards.

Background

Luther Simjian was a natural inventor who garnered a great number of patents during his lifetime. He developed a self-posing portrait camera that was licensed and manufactured for department stores. He also invented the Range Estimation Trainer, designed to train pilots during World War II.

The idea of an ATM was not new when the first modern machine was installed as a walk-up outside a New York bank in the early 1970s. A remote machine that would facilitate financial transactions was the brainchild of Turkish-born American Luther Simjian back in 1939. At that time, the company that is now Citicorp reluctantly decided to give Simjian's idea a trial run, but the world was not ready for such mechanical convenience at the time.

"It seems the only people using the machines," Simjian wrote, "were a small number of prostitutes and gamblers who didn't want to deal with tellers face to face." Simjian would never cash in on his invention—it was removed after only six months.

A few decades later, the world had grown a little more tech-savvy. While waiting in line at a bank, Don Wetzel "reconceived" the ATM. Wetzel and the two colleagues listed on his patent application worked for Docutel at the time, a company that developed automated baggage-handling equipment.

The three men developed plans for an off-line machine that would record and store information about transactions through coded information on what were the first ATM cards. But the machine was not linked by computer to the central bank. Until on-line systems were developed, banks were very selective about which clients could enjoy ATM privileges.

Today, ATMs are significantly more advanced and can perform a variety of banking functions in addition to dispensing cash. With the machines on nearly every street corner throughout the world, stepping inside of a bank has become, for many, a thing of the past.

How It Works

1. An encoded plastic card is inserted into or "swiped through" a code-actuated automatic currency machine, which contains code-scrambling technology.

2. The user enters a personal identification code that is checked against the coded information stored in the machine.

3. Once verified, an automated dispensing apparatus reacts to the user's commands.

FIG. I

FIG. 4

INVENTORS:
THOMAS R. BARNES
GEORGE R. CHASTAIN
DON C. WETZEL

Richards, Harris & Hubbard
ATTORNEYS

In the Inventors' Words

"A currency dispenser automatically delivers a medium of exchange in packets in response to a coded credit card presented thereto. The coded credit card is presented to the currency dispenser and an initial check is made to determine if the card has the proper format. After checking the credit card format, coded information thereon is evaluated to check the user's identity prior to authorizing him to receive cash from the machine. When several additional checks of the credit card code have been completed, the old code is removed and substituted with a new code. The new code contains the same information as the old but updated to reflect an additional currency dispensing transaction. Both the original code and the updated code are scrambled in accordance with a changing key. Scrambling the credit card code after each use thereof minimizes the chance of unauthorized use of the currency dispenser. When the check of the credit card code indicated the user is entitled to receive the amount of currency he has selected, a storage container for the packets of currency transports the required number of packets by a positive feed drive to a cash drawer. The cash drawer opens to a detent position…allow the customer to then move the drawer to a fully open position to remove his currency. Upon release of the cash drawer, it returns to a partially opened position from which it automatically closes after a preset time limit."

A velocipede in 1868

Bicycle

Patent Name: Improvement in Velocipedes
Patent Number: 59,915
Patent Date: November 20, 1866
Inventor: Pierre Lallement, of Paris, France

What It Does

The bicycle provides a means of personal street conveyance via a pedal-operated, wheeled vehicle for one person.

Background

For numerous generations, the bicycle has been providing sheer merriment to people of all ages. Velocipedes, as they were called in France, or bone-shakers in England, have been around for quite some time. Many individuals from a variety of countries were involved in the creation of the bike, dating as far back as the seventeenth century, long before Pierre Lallement received a U.S. patent for his in 1866. By 1895, so many patents were being awarded for variations on and modifications to the bicycle that the U.S. Patent Office opened a separate department for such inventions. One popular model developed in the 1870s was the "penny-farthing," which featured a high seat situated above an enormous front wheel, about four times the diameter of the back wheel.

While there remains a great deal of dispute over the origin of the bicycle, Pierre Lallement is accepted by many as its inventor, the father of the modern bicycle. In the early 1860s, Lallement worked for a company that produced wheelchairs and baby carriages, but he started toying with the idea of making his own products. Realizing others were working on similar inventions, Lallement traveled to America with some parts and gained access to a workshop in Connecticut. Here, he put the finishing touches—pedals—on what would become the first bicycle-related patent in the United States.

Wheels were made of solid rubber until 1887, when John Boyd Dunlop made water-filled tires out of a garden hose, for a tricycle. He later used air, and pneumatic tires soon became the norm.

The first women's bicycle frame was designed by one of the Duryea brothers, the pioneers who built the first American automobile. The two siblings were originally bicycle makers. The women's frame was angled so as to allow the more petite riders to saddle more easily.

Built by a Dutch cycling organization, the longest bicycle is over 117 feet long. One person must steer in the front and the other must pedal from the back.

Today, throughout the world, bicycles provide a popular form of transportation, recreation, and sport. Appropriately enough, Lallement's home country hosts the most celebrated bike race in the world: the Tour de France.

This model accompanied another "velocipede" patent application, an early tricycle, awarded to E. J. Blood in 1891 (patent number 211,959).
Rothschild Petersen Patent Model Museum

P. LALLEMENT.
VELOCIPEDE.

No. 59,915.   Patented Nov. 20, 1866.

Fig.1

Fig.2

Fig.3

Witnesses

Inventor:

How It Works

1. A frame (C) connects the axles of two wheels (A, B).
2. The arms of the front wheel are arranged on a pivot that is controlled by handlebars (D).
3. Attached at the front axle is a pair of cranks (E) that connect to pedals (F).
4. A rider straddles a seat and begins pedaling, propelling forward while balancing his weight on the frame.
5. Braking would come a little later, but the rider did have some influence on slowing the rate of the pedal revolutions.

In the Inventor's Words

"My invention consists in the arrangement of two wheels, the one directly in front of the other, combined with a mechanism for driving the wheels, and an arrangement for guiding; which arrangement also enables the rider to balance himself on the two wheels."

Breech-Loading Ordnance

Patent Name: Improvement in Fire-Arms
Patent Number: 30,045
Patent Date: September 18, 1860
Inventor: Charles F. Brown, of Warren, Rhode Island

What It Does

Breech-loading cannons were not the most popular American Civil War–era firearms, but they did introduce novel ideas in gun loading that would eventually speed up the firing action of guns. This one could be likened to an elderly ancestor of the semiautomatic machine gun.

Background

Between the years 1861 and 1865, some 10,000 sites served as settings for the American Civil War. More than 600,000 soldiers fought in it, and 2 percent of the entire American population died in its battles. In 1863, the war took a turn for the North when Union soldiers defeated Confederates at the Battle of Gettysburg, Pennsylvania. On January 31, 1865, Congress succeeded in passing the Thirteenth Amendment to the United States Constitution, abolishing slavery.

Before and during the bloody years of the Civil War, gun makers were scrambling around trying to secure government contracts. Dozens of rifled cannons such as this one were made, and there are many classifications for them. Breech-loading (loaded through the base) and muzzle-loading were two types, but most Civil War cannons were muzzle-loading, as the breech-loaders were not yet perfected for battle. The breech-loading cannon was certainly ahead of its time. This Charles Brown invention, for example, is designed to self-load and fire as it is pushed toward enemy lines, reducing the time and labor between each explosion.

A breech-loading cannon in Richmond, Virginia

The original patent model
*Rothschild Petersen Patent Model
Museum*

C. F. BROWN.

Breech Loading Cannon.

No. 3,211.

Reissued Nov. 24, 1868.

Fig. 3.

Fig. 2.

Fig. 1.

Witnesses:

Inventor:

How It Works

1. A handle (H) is constructed to attach to the axles of two wheels (D) toward the front of the gun.
2. A third wheel (J) is situated at the back, immediately juxtaposed to the breech of the cannon.
3. As the gun is moved, the third wheel revolves its axle shaft (I), evident in figure 2.
4. The revolving shaft causes a cam (K) to drive forward a breech-pin (b) beyond apertures (f, f).
5. A number of cartridges may be arranged in a magazine above the apertures, which will drop in one by one as the breech-pin retreats and allows access to the apertures.
6. The gun can also be fired in a stationary position when the rear wheel is raised off of the ground and a crank (as shown in the model photo) is attached to one end of the shaft to rotate the wheel and move the cam.

In the Inventor's Words

"The object of my invention is to enable a piece of ordnance to be fired repeatedly by the act of its being drawn over the ground, so that it may be made very effective in advancing to meet or pursue or in retiring from before an enemy."

Boys sorting coal coming out of the breaker in Scranton, PA

Coal Breaker

Patent Name: Improvement in Coal-Breakers
Patent Number: 219,773
Patent Date: September 16, 1879
Inventor: Philip Henry Sharp, of Wilkes-Barre, Pennsylvania

What It Does

The coal breaker is a large machine that is used in coal mining operations to break up the chunks of raw coal removed from the mines into pieces of a more manageable size.

Background

Black lung disease kills more than one thousand coal-mine workers annually.

A finite natural resource found beneath the Earth's surface, coal is a fossil fuel that contains amorphous carbon with various organic and some inorganic compounds. In its heyday, coal fueled the Industrial Revolution in both Europe and America. It powered westward expansion on the railroad in the United States, and trade and commerce on steamships on waters throughout the world. Coal was, and still is, used to generate electricity, to heat homes, and to fire furnaces used to smelt iron ore for iron and steel.

In the mid-1800s, Pennsylvania's Wyoming Valley played a major role in the "black gold" rush. Veins of anthracite coal stretched everywhere beneath the ground in the surrounding hills, and shafts were dug left and right. Wilkes-Barre became a major railroad stop, a huge coal-mining center, and a mecca for tens of thousands of Welsh, German, English, Irish, and Russian immigrants in need of work.

These poor immigrants were exploited by the mining companies. Young boys were employed to operate coal breakers that sorted the mined coal from rock and simultaneously crushed the coal into manageable pieces. While the breaker was at the bottom rung and the miner was at the top, life in the shafts was by no means glorious. Working conditions were deplorable, and accidental death was an occupational hazard and a common reality.

The original patent model
Rothschild Petersen Patent Model Museum

There are four basic types of coal:

Anthracite is a hard coal used for heating and generating electricity.

Bituminous is a soft coal commonly used for generating electricity. As a fuel, it is not as efficient as anthracite.

Subbituminous is a dull black coal found beneath the layer of bituminous coal and is used for generating electricity and heat.

Lignite is another hard coal. It has a higher ash and moisture content, so its value is not as high as anthracite's.

P. H. SHARP.
Coal-Breaker.

No. 219,773.   Patented Sept. 16, 1879.

Fig 1

Fig 2

Witnesses;
A. C. Faggart
J. A. M. Connell

Inventor;
Philip Henry Sharp.

How It Works

1. Set on a frame (b), two transversely secured plates (a, a) inclined toward each other feed into a hopper (x) through an opening toward their lower ends.
2. Corresponding horizontal apertures line each of the plates, but are on separate vertical planes.
3. Parallel with the plates, two shafts (c, c) are mounted in bearings (c'); each shaft having firmly secured upon it a series of eccentrics (d'), provided with straps (d).
4. Picks or breakers (e) are fitted to straps and secured by a key or setscrew.
5. The reciprocating picks project through the plate apertures, and act upon the coal as it is fed through the hopper.

In the Inventor's Words

"As ordinarily constructed coal-breaking machines are composed, substantially, of two parallel rollers, the peripheries of which are studded with teeth, which break up the lumps of coal fed to and passing between the rollers. Such construction is objectionable in the regard that the crushing action of the rollers pulverizes the coal to a considerable extent and entails a corresponding amount of waste, and, with a view to the avoidance of such result, the rollers have in some cases been dispensed with and the coal broken by means of vibrating plates, each carrying a series of picks or breakers operating within a stationary hopper."

Cotton Gin

Patent Name: Cotton Gin
Patent Number: 72X
Patent Date: March 14, 1794
Inventor: Eli Whitney, Philadelphia, Pennsylvania

What It Does

Mechanically cleans and separates out the seeds from cotton, so that it can be further processed for a variety of uses.

Background

One of the earliest patents in the United States was awarded to Eli Whitney for his now-famous cotton gin in 1794. It is rumored that Whitney got the idea for the device when he saw a bird fly against the bars of its cage and lose feathers in the process; others claim his inspiration came from witnessing a cat trying to catch chickens in a wire coop. He envisioned that the cotton gin would streamline work and revive flagging profits for the American South's declining cotton industry—and that he might gain a share of that profit.

Along with his business partner, Phineas Miller, Whitney offered to gin the cotton for farmers throughout the South, on the condition that he got to keep a share amounting to two-fifths of the profit. Instead of being heroes, though, Whitney and Miller inspired bitter feelings in the farmers, who thought their fees unfair. Farmers elected instead to make their own versions of the cotton gin, and the South prospered, while little was done to protect Whitney's patent.

Whitney went north, where he had better luck making muskets. Although he didn't profit from his cotton gin, Whitney's reputation as a shrewd inventor was well established and he was able to secure a contract from the U.S. government to produce one thousand rifles. In meeting his end of the deal, Whitney honed a method for factory-producing and assembling various gun parts, much to the dismay of skilled gunsmiths and other experts in the craft.

Equality! Or Yankee Ingenuity? As the availability of cotton increased, so did the monetary value of slave labor. Cotton was farmed in the South and milled and sold in the North, and resentment grew as the North began collecting taxes on the sales. Southern farmers started doing their business overseas, prompting further enmity between North and South. Some skeptical historians believe the slave controversy was no more than an outgrowth of disputes over the market of raw goods and that the moral issue was a veiled pretense by the Union to maintain an economic upper hand.

A wood cut of the cotton gin created from the original model

How It Works
1. Raw cotton is delivered by means of a shaft and conveyer into the main apparatus.
2. Within the apparatus, a drum cylinder rotates past a sieve, pressing against the incoming cotton.
3. The sieve separates the seed from the lint.
4. Hook-shaped claws on the drum catch the lint.
5. A rotating brush sweeps the cotton from the claws.

In the Inventor's Words
"This machine may be turned by horse or water with the greatest ease. It requires no other attendance than putting the Cotton into the hopper with a basket or fork, narrowing the hopper when necessary, and letting out the seeds after they are clean."

Door Lock

Patent Name: Door Lock
Patent Number: 3,630
Patent Date: June 13, 1844
Inventor: Linus Yale, of Springfield, Massachusetts

What It Does
Secures a door so that entry is possible only with a key, providing security and privacy.

Background
As a young man, Linus Yale wanted to be a portrait painter, but his artistic energies soon gave way to solving mechanical problems. He decided to follow in his father's footsteps and become a lock maker. This simple door lock came before

Yale achieved fame and fortune for his 1851 Infallible Bank Lock. This success allowed him to set up his own shop and continue fashioning the kind of locks that would secure banks from robbers, homes from intruders, and bathroom users from unexpected surprise entrances.

How It Works

Linus Yale, Jr.

1. A tumbler (E) is a simple cylinder of the same diameter as the cylindrical socket in which it is placed and made to revolve.
2. It is pierced in its center with a cylindrical cavity to admit a key, and perforated around its periphery with the same number of radial apertures as there are in the circular rim (C) inside the front plate of the lockbox.
3. On the revolving cylinder's perforations are placed pistons (F) that drive another set of pistons (D) back out of the apertures in the tumbler until their ends coincide with the outer periphery of the tumbler and inner periphery of the rim.
4. In this position, the tumbler can be turned for locking and unlocking the bolt.
5. A key (K) is a cylinder containing as many wedge-shaped cavities or grooves (x) in its periphery as there are pistons.
6. The key also contains a cog (k), which enters a corresponding cavity or notch (n) made in the tumbler.

"The improvement consists in having a cylinder or circular rim C on the inside of the front plate of the box fastened thereto in any convenient way, or cast therewith, forming a cylindrical socket, said cylinder or circular rim being pierced from the outer to the inner circumference with round apertures on lines radiating from the center to the periphery, in which apertures are placed cylindrical pistons D, which pass through said apertures and enter corresponding apertures in a revolving cylinder or tumbler E, (having a hub or arm) that turns in said cylindrical socket for locking the bolt as hereafter described, said pistons being kept pressed inward toward the center (against other pistons F hereafter described) by springs G attached to the front plate of the box."

An early indoor flush toilet kept odors trapped inside. The next challenge was a cosmetic one.

Flush Toilet

Patent Name: Detachable Flush-Rim Fixture
Patent Number: 1,107,515
Patent Date: August 18, 1914
Inventor: Philip Haas, of Dayton, Ohio

What It Does

The flush toilet is an indoor sanitation device for removing human waste. It also serves many people as a quiet place to enjoy reading material, such as the book you are now holding.

Background

Thomas Crapper was an English plumber who in the mid-1800s advanced indoor toilet technology. Despite the coincidence of his name, he did not invent the contraption.

The origins of the commode extend far back in history. The technology that employs running water to remove human waste is much younger. A pivotal moment in the development of the toilet came in 1775, when Englishman Alexander Cummings invented a water closet that trapped water in a U-shaped section of pipe to block the noxious odors below in what was then called a "stink trap." A valve on a hose connected to the bottom of a basin was turned to flush the debris down—and the flush toilet was in its infancy.

The New World was a bit slow in adopting this technology. Several decades passed before the first American received patents for water closets. But by the end of the 1800s, great and numerous improvements would appear. This patent is among hundreds of toilet inventions developed in the United States in the first three decades of the twentieth century. It describes a detachable fixture with apertures used to manipulate water from its source and keep the bowl clean.

Eventually, toilet assemblies would include tanks and water-conserving siphons, which rely on atmospheric pressure to push water from a reservoir, over a barrier, and out the other end.

How It Works

1. Attached to a water supply pipe, the flush rim fixture is made to fit snugly within the rear of the bowl's lip.
2. The fixture includes a series of perforations (11) that project water downward over the interior rear surface of the bowl.
3. At the fixture's extreme ends, additional orifices (12) are made to direct jets of water in opposite directions about the periphery of the bowl.

In the Inventor's Words

"Heretofore closet bowls of the type herein shown, have usually been constructed with a water conduit extending about the periphery of the bowl beneath the inturned flange thereof. This conduit is usually provided with perforations, which become clogged by deposit or sediment. Such flush conduits being located beneath the level of the bowl rim are unsanitary, since any stoppage of the soil pipe which might cause the bowl to fill or overflow, would result in the contents of the bowl filling such flush conduits and backing into the supply pipe, thus perhaps contaminating the water supply of the house."

Frozen Food

Patent Name: Method of Preparing Food Products
Patent Number: 1,773,079
Patent Date: August 30, 1930
Inventor: Clarence Birdseye, of Gloucester, Massachusetts

What It Does

This means of packaging and freezing fresh food insures that it remains stable over long periods and retains its flavor, texture, and color when defrosted for use.

Background

While intensive research was being conducted to develop a better refrigeration system, which led to the 1928 invention of Freon, Clarence Birdseye was applying for patents related to preparing and packaging frozen foods. The scales were already tipping in his favor.

It makes perfect sense that Birdseye became the father of frozen foods. After quitting college, where he majored in biology, he worked for the U.S. government as a naturalist in the Arctic. Birdseye saw how the native people capitalized on their environment. They kept freshly caught fish in barrels of frozen seawater, and the fish still tasted fresh when cooked long after it was caught. It froze so quickly in the subzero temperatures, he concluded, that the cellular structure of the food remained intact.

Birdseye went to New York to work on replicating this natural phenomenon with an electric fan, buckets of brine, and lots of ice. Eventually, he perfected a system of packing and flash-freezing fresh food that has inextricably linked his name to what is now a multi-billion-dollar frozen food industry.

Birdseye, seen here on the assembly line, helped usher in the era of modern convenience foods.

By the end of the 1950s, frozen food sales had exceeded $1 billion; more than 60 percent of all retail grocers were stocking frozen foods; and frozen foods were being used on airline flights.

How It Works

Spinach was the first frozen vegetable to be sold.

1. Food is consolidated into a compact form.
2. The form is encased in the wrapper or carton in which it will be later sold.
3. On a system of conveyer belts, the package is further compacted by means of metal plates.
4. Simultaneously, the metal plates—maintained at temperatures of between –20° and –50° Fahrenheit—help facilitate the quick-freeze of the product as it travels through a freezing chamber.

In the Inventor's Words

"My invention relates to a method of treating food products by refrigerating the same, preferably by 'quick' freezing the product into a frozen block, in which the pristine qualities and flavors of the product are retained for a substantial period after the block has been thawed."

A poster advertising "escape artist" Harry Houdini

Handcuffs

Patent Name: Improvement in Handcuffs
Patent Number: 200,950
Patent Date: March 5, 1877
Inventor: John J. Tower and Henry W. Kahlke, of Brooklyn, New York; assigned by John J. Tower

What It Does   Handcuffs are used primarily as a portable temporary restraining mechanism by law enforcement officers.

Background   Manacles and restraint systems have been around as long as war and crime have existed. In ancient Egypt, Jewish slaves were shackled together, often in pairs. In colonial America, petty crimes were punishable by putting the perpetrator in pillory—wooden frameworks—in the public square. For ages, people have been trying to perfect the mobile temporary restraints known as handcuffs. Around the turn of the twentieth century, the name Tower dominated the handcuff industry. John Tower took the adjustable ratchet principle from another handcuff maker of the day, W. V. Adams, who patented an earlier restraining device in 1862. This new generation of cuffs restrained criminals of all sizes. The handcuff efficiency likely annoyed apprehended criminals who remembered an earlier era of simple metal ring cuffs, more easily escapable. Tower cuffs remained the standard until the 1940s.

How It Works   1. In Figure 1, the notched segment (a) connects by joint (b) with a radius-bar (c).
2. A link (d) for the chain is also connected to the joint.

3. The hollow-bodied radius-bar contains the lock case.
4. In Figure 2, two independent spring catches (e, e[1]) lock onto the notches, thereby securing the lock more firmly and making "lock-picking" more difficult; the normal cuffs of the day had only one spring catch.
5. In figure 1, a tumbler (1) swings on a stud (3) and has notched ends.
6. A fence or stud (o) on the bolt (h) prevents the key from withdrawing the bolt until the tumbler is in position for the stud to pass into the tumbler notches.
7. In Figure 2, a cylinder (r) is grooved longitudinally to receive the plate key and be turned. In locked position, the cylinder is held in place by a cap plate (u).
8. The cap plate is secured in place by studs (3, 4), and its position against the lock case itself.

In the Inventors' Words "Handcuffs have been made with a segmental notched loop and a radius-bar containing the lock or catch. Our improvement is made to avoid the risk of the lock or catch being opened, either by an instrument introduced through the keyhole, or by a quill or thin piece of metal inserted between the segmental bar and the radius-bar."

Handle Guide

Patent Name: Handle Guide
Patent Number: 2,377,745
Patent Date: June 5, 1945
Inventor: Emile Belanger, of Drummondville, Quebec, Canada

What It Does

The handle guide provides an efficient means of holding, and thereby operating, hand tools and represents an early understanding of ergonomics.

Background

A tool should be crafted to fit as harmoniously as possible with the hand. This is the defining principle behind Belanger's handle guide patent. Belanger's invention is a series of strategically placed grooves and ridges on a tool handle, meant to conform to the hand. Many computer keyboards now adhere to the same principle; the handle guide is a close cousin to modern-day ergonomics.

Emile Belanger's invention may have been simple, but he himself was not. He was seen in public only in full suit and tie—no matter the season or occasion. The rigidity of Belanger's dress code seemed tame in comparison with his insatiable hankering for fresh seafood. According to his granddaughter, the inventor's wife suffered a fainting spell one evening when, upon returning home, she discovered the evening's meal swimming in the bathtub in the form of a giant eel.

In spite of his peculiarities, Belanger was a straightforward thinker who enjoyed his business of building houses in Quebec. He was quick, efficient, and had a handle on tools.

How It Works

1. The handle (A) features a gripping portion, rounded at the top and decreasing in diameter towards the center.
2. Below the center is a ridge (1) extending out to about the same diameter as the top.

3. Below the ridge is a depression, which curves back in to about the same diameter as the center of the handle.
4. This is followed by a smaller ridge (3) which tapers out into triangular facets (4).
5. The first ridge (1) provides a firm hold for the thumb, and six cavities (5) allow for the placement of the fingers (other than the index finger).
6. The index finger, which really guides the motion of the tool, can extend farther along the tool to rest firmly on any of the inner cavities (6).

Today, the ergonomic principles employed in this patent have become the foundations for both a science and an entire industry. The science of ergonomics considers the correlation between mechanical properties of human tissue and the response of the tissues to mechanical stress. In other words, building a more comfortable desk chair or a computer mouse that doesn't cause carpal tunnel syndrome takes advantage of the study of ergonomics.

June 5, 1945.   E. BELANGER   2,377,745
                        HANDLE GUIDE
                     Filed April 18, 1944

FIG. 1.   FIG. 2.

E. Belanger
Inventor

By: Kimmel & Crowell
Attorneys

In the Inventor's Words   "The operator should have proper means to guide his hand and fingers so that when he is operating the tool there should be the least variance in its motion. I have devised such a guide and in using same the operator is also saved the strain on his eyes due to the fact that he need not have his sight fixed permanently on the operation to detect or divert any faulty motion."

"It is important to note the relative alignment of these cavities for the thumb and index fingers. The division line between any two cavities (6) is opposite the centre of a cavity (5) in ridge 1. This alignment is made to correspond to the normal position the fingers should take on a handle when a tool is being properly operated."

John Deere Plow

Patent Name: Improvement in Molds for Casting Steel Plows and Other Articles
Patent Number: 41,203
Patent Date: January 12, 1864
Inventor: John Deere, of Moline, Illinois

What It Does

John Deere patented the process he used for making the kind of plow that earned him a great reputation in his day, and an even greater legacy in the company that now bears his name. His secret to success was in the way he cast molds, which resulted in an extremely efficient plow blade.

Background

Deere & Company now employs about 57,000 people and does business in more than 160 countries across the globe. Its product line has expanded considerably to include everything from riding mowers and skid steer loaders to forestry excavators and dump trucks. In 2016, Deere & Company's annual sales and revenues were $26 billion. The company marked the 200th birthday of company founder John Deere on February 7, 2004.

The John Deere story began about forty years prior to the date of this patent, when Deere honed his skills as a blacksmith in Vermont. Tales of opportunities in the "West"—in this case, Illinois—eventually lured Deere to move there; but all was not bliss. Farmers were discouraged by the way the rich soil, in contrast to the soils back east, clung to the plows and had to be scraped at frequent intervals. Deere made it his mission to solve the problem, which he did in 1837 by using steel to craft a polished and uniquely shaped plowshare that fastened to the moldboard—the curved plate that cuts and turns the earth. Until then, plows had been made only of wood or cast iron.

But success was not instant for John Deere. He made one plow in 1837. He made two in 1838. He made ten in 1839. By 1840, he still considered himself a blacksmith, according to his 1840 census form. A decade later, though, he registered himself as "plow maker," and by the year 1858, Deere was producing one thousand plows annually. In 1864, he patented his unique idea for molding plows. Four years and a few patents later, John Deere's company became incorporated. Today, Deere & Company is one of the oldest industrial companies in the United States.

John Deere ad from 1852

How It Works

A mold is formed, coated with black lead, and baked.

In the Inventor's Words "I form a mold by means of a suitable pattern out of a dry sand in the usual way, and after the mold is formed I coat its inner surface with black-lead (plumbago) which may be moistened by water having a little fire-clay dissolved in it. The sand also used in forming the mold may be tempered with fire-clay water. This fire-clay water has a tendency to render the sand and black-lead or plumbago rather more adhesive than they would be if water alone were used. After the mold is formed and coated internally with the black-lead or plumbago, it is baked, say, for about twelve hours, or until all moisture is thoroughly expelled from it, and when thus baked and thoroughly dried the black-lead or plumbago will form a hard crust, coat, or glazing for the inner surface of the mold, so as to prevent the melted steel from coming in contact with the sand. The latter therefore cannot be fused and cut away by the melted steel, and will form or produce a perfect casting, or one with a smooth surface and free from holes or air-cells. This mold may, before being baked and dried, be perforated with vent-holes to admit of the escape of air and gases. I would further remark that the mold is used but for one casting only, it being, after once used, broken up or pulverized and the same ingredients used for the formation of another mold."

Cats are very meticulous, constantly cleaning themselves. Their rough-surfaced tongues are designed to lick off any scents that may attract insects or predators.

Kitty Litter

Patent Name: Cat Box Filler with Incorporated Pesticide
Patent Number: 4,664,064
Patent Date: May 12, 1987
Inventor: Edward H. Lowe, of Cassopolis, Michigan

"The interesting thing about being a mother is that is that everyone wants pets, but no one but me cleans the kitty litter." —Meryl Streep

What It Does

Kitty (or cat) litter is an absorbent, easy-to-replace substance resembling sand, used in a box or bin as a place for house cats to do their business. The litter keeps odors and messes to a minimum, and cats seem to appreciate having their own neat and tidy toilet facility.

Background

With the introduction of commercial cat litter products, the cat became more of an indoor pet.

Keeping your cat indoors can:

- Reduce the spread of diseases, such as rabies.
- Save birds, mice, and lizards, which all naturally control insect populations.
- Reduce its chance of getting mauled by the neighbor's German shepherd or tortured by the local delinquent.

This patent is but one of dozens awarded to Ed Lowe, who has earned the odd designation of "the father of kitty litter." Lowe's own father, Henry, was a small business owner who sold industrial absorbents in 1948. Fresh out of the Navy, Ed joined Dad in the biz and tried in vain to sell farmers a new "fuller's earth" product as a nesting material made from partially oven-fired balls of dried clay. While local farmers did not exhibit much interest, Ed's next-door neighbor did. In response to her cat-related woes, Ed suggested using the absorbent material. She was very impressed by the results, and kitty litter was born.

Until kitty litter marked its territory in the commercial market and established the beginnings of a new industry, cat owners commonly used shredded newspaper, dirt, or sand for in-home litter. In the meantime, Lowe became an ambitious traveling salesman and hit the road aggressively hawking his kitty litter. As his product began to absorb public attention and sales, Lowe established Edward Lowe Industries, Inc. and concentrated his energies on making improvements, such as added ingredients to kill odor-causing bacteria or to create moisture-activated clumping action for easier cleaning. This patent represents a litter that contains pest-controlling chemicals, helping to keep your cat flea-free while it makes pee. The kitty litter king died in 1995.

How It Works

1. The pesticide may be either vapor-producing or contact-actuated.
2. A contact-actuated pesticide is impregnated into clay, or any solid, nontoxic organic polymer.
3. The product should be packaged either directly with the cat box filler or in a separate pouch to ensure containment.
4. When the cat scratches in its litter box, it releases the pest-control chemicals.
5. Examples of contact-actuated pesticides are heptenophos, carbaryl, propoxur, resmethrin, naled, trichlorfon, tetrachlorvinphos, diazinon, and fenchlorphos.

In the Inventor's Words

"Pest control in domestic animals has long been a major concern of pet owners. Previously, adequate pest control has been achieved by the use of a pesticide-impregnated collar worn by the animal. The major drawback of these collars is that they must be worn constantly by the animal to be effective. This constant wear increases the risk of pet poisoning and presents a danger of accidental strangulation in cats.

"The pest control of this invention involves the incorporation of a pesticide into granules which are dispersed in the cat litter box, when the cat uses the litter box, the pesticides granules bathe the cat in vapor and prevent the outbreak of fleas and other pests. Because the granules are dispersed within the cat box filler, pest control is achieved in the litter box as well as on the cat itself."

The rotary blade lawn mower was heavy and hard to push uphill, but at least it was quiet.

Lawn Mower

Patent Name: Lawn Mower
Patent Number: 624,749
Patent Date: May 9, 1899
Inventor: John Albert Burr, of Agawam, Massachusetts

What It Does

This simple machine keeps a lawn trimmed uniformly to the desired height with a minimum of physical labor, by means of blades that rotate as the device is propelled forward.

Background

An alternative to more costly motor sports, the first British Grand Prix Lawn Mower Races were held in 1973, with separate races for run-behind mowers, towed seat mowers, and tractor seat mowers. Today, the fastest lawn mower goes more than 133 mph.

Edwin Budding of Gloucestershire, England, patented the first "machine for mowing lawns" in 1830. Budding's design was based on carpet-trimming tools, with multiple blades attached side by side around a shaft.

Connecticut resident Amariah Hills received the first U.S. patent for a reel lawn mower in 1868; the growing expanse of suburbs along the East Coast of the United States was creating demand for an efficient machine of this sort.

The first rotary blade lawn mower was created by fellow Connecticut resident John Albert Burr in 1899. African American inventor Burr added some significant components to the device that set a new standard for a safer and better lawn mower. He included a set of three curved rotary shears that prevented the freshly clipped grass from clogging the gears, and traction wheels set within the reach of a longer blade. This design allowed the mower to access to the grass while reducing the overlap caused by previous wheel positions.

How It Works
1. Wide traction wheels (A) are set at the inside ends of an axle shaft (B).
2. Cheek plates (C) secured on the outer edge, the hub, of each tire support an outward-extending central shaft (f), the fixed bar of the cutter (k), and a countershaft (i).
3. Loosely mounted on the axle shaft is a gearwheel (O), which engages a ratchet wheel that synchronizes the rotation of the blade with the motion of the traction wheels.
4. As the operator pushes the device forward, the blade rotates and shears the grass in a uniform swath.

In the Inventor's Words

"By making the side cheek-plates with the forward extension members decidedly out-forwardly offset beyond the outer faces of the traction-wheels provides for the mounting between the forward ends of the cheeks of a rotary cutter having a length quite a little exceeding the distance between the outer faces of the traction wheels, and hence it becomes possible to cut the grass more closely to a building or other object than would be possible were the ends of the rotary cutter disposed within the planes coincident with the outer faces of the said wheels, and it will be seen that in the use of the machine its every pass will cut the grass in a path wider than that comprised between the traction-wheels, so that the disadvantage of having one of the wheels bed down the grass just outside of the cut path or swath at every pass is overcome."

Microwave

Patent Name: Method of Treating Foodstuffs
Patent Number: 2,495,429
Patent Date: January 24, 1950
Inventor: Percy L. Spencer, of West Newton, Massachusetts

What It Does

Cooks food quickly by microwave radiation.

Background

In 1946, Percy Spencer, a self-taught electronics wizard who never graduated grammar school, received U.S. patent 2,408,235 for the development of a high-efficiency magnetron. Magnetron tubes employ heated cathodes to generate electrons that are affected by electromagnetic energy and produce microwave radiation. Microwave radiation is used both in radar and in the "method of treating foodstuffs" that Spencer invented and patented four years later.

While working for the Raytheon Manufacturing Company, Spencer was running tests on one of the new magnetron tubes when he noticed the candy bar in his pocket begin to melt. Employing his natural inventor instincts, he decided to place some uncooked popcorn in front of the magnetron tube. To his delight, the kernels began popping. (This is still a favorite application of microwave technology.)

Engineers at Raytheon went to work on Spencer's discovery. The first commercial microwave oven was introduced in 1947. It was 750 pounds and five and a half feet tall. Twenty years and many modifications later, the first counter-top domestic microwave oven hit the market. By 1975, microwaves were selling faster than conventional gas-range ovens.

How It Works

Many urban legends followed the introduction of the microwave. For example: An elderly lady who bred Persian cats used to wash her cat, towel it dry, and then warm it in her electric oven. Her son brought her a brand-new microwave oven for Christmas. On the day of the next cat show, not understanding the technology, she washed her prize-winning Persian cat and popped it into the microwave for a few seconds. There was no time for a meow. The cat exploded the instant the oven was switched on. Poodles, babies, and exploding glasses of boiling water have all had their turn in the urban folktales involving microwaves.

1. Electron-discharging cathodes (14) housed within oscillators are joined by conductors (20, 21).
2. These conductors are in turn connected by another conductor (22), a center tap on a winding (17) of the transformer (18).
3. Coaxial transmission lines (24, 25) alternately deliver hyperfrequency energy to a wave guide (23).
4. The wave guide is directed at a piece of food on a speed-adjustable conveyor (28).

In the Inventor's Words

Spencer held 150 patents during his life. In 1999, he was inducted into the National Inventors Hall of Fame.

"Such energy has been used before for this purpose, but the frequencies employed have been relatively low, for example, not over 50 mega-cycles. I have found that at frequencies of this order of magnitude, the energy necessarily expended in order to generate sufficient heat to satisfactorily cook the foodstuff is much too high to permit the practical use of the process. I have further found, however, that this disadvantage may be eliminated by employing wave lengths falling in the microwave region of the electromagnetic spectrum, for example, wave lengths of the order of 10 centimeters or less. By so doing, the wave length of the energy becomes comparable to the average dimension of the foodstuff to be cooked, and as a result, the heat generated in the foodstuff becomes intense, the energy expended becomes a minimum, and the entire process becomes efficient and commercially feasible."

Microwave **199**

Pasteurization
plant

Pasteurization

Patent Name: Improvement in Brewing Beer and Ale
Patent Number: 135,245
Patent Date: January 28, 1873
Inventor: Louis Pasteur, of Paris, France

What It Does

This heat-based process, named for its inventor, kills bacteria in perishable liquids such as beer and milk, and stabilizes and preserves them without significantly compromising taste or quality.

Background

Louis Pasteur was a brilliant scientist whose work is memorialized by the semisterilization process named in his honor. The French scientist disproved the long-held theory of spontaneous generation, which maintained that microorganisms arise in decaying biological matter. Pasteur believed that microorganisms existed everywhere, that they were not symptoms of decay but rather the cause, and that the contamination of food and beverages by bacterial microbes could be prevented through a specialized heat treatment that minimized exposure to microbe-carrying air, yet did not destroy the food.

Pasteur put his theories into practice by coming up with an improved process of fermentation. Previously, fermentation of sugar by yeast had yielded the desired alcoholic product, but also some undesirable by-products. Pasteur believed these were related to particular microbes, or "germs," which were responsible for fermentation, but could be separated out and treated for a better result. He also demonstrated that particular germs were responsible for many types of diseases, and in their weakened forms, some microbes could be used for immunization purposes.

Three of the inventor's five children had died from typhoid fever, and Pasteur was determined to unlock the mysterious causes of disease. He was quite successful, too. Pasteur found that rabies was transmitted by viruses that had previously been undetected because they were invisible even under a microscope. He subsequently developed a vaccination against rabies. Altogether, Pasteur's work revolutionized food sterilization and disease treatment, and he single-handedly initiated various new fields of medical and scientific study.

Louis Pasteur

How It Works

Why does draft beer taste fresher? Because most commercial breweries bypass pasteurization of keg beer (which can reduce flavor) and insist on keeping it properly refrigerated.

Today's large commercial breweries still use pasteurization. Beer is aged in tanks, filtered, artificially carbonated, and then pasteurized, which preserves the beer's freshness by eliminating spoilage organisms or residual brewing yeast that can compromise the quality of the beer. The process Pasteur developed also opened the doors for bottling, allowing for at-home enjoyment of the beverage.

1. In the pasteurization of beer, boiling-hot wort—unfermented malt infusion—is introduced into a tank (A).
2. From gas supply tanks (M), carbonic-acid gas is projected into the tank through piping (w) to remove air from the wort.
3. A water supply pipe (E) delivers spray through nozzles (P) to the outside of the tanks to cool the wort.
4. The wort is left or delivered into another vessel, where it ferments.

In the Inventor's Words

"My invention has for its object to produce a better quality and greater quantity of beer from the same quantity and quality of wort, and to afford a beer which shall also embody the quality of greater degree of unalterableness during time and changes of climate &c., in transportation and use; and to these ends my invention consists in expelling the air from the boiled wort while confined in a closed vessel or closed vessels, and then cooling it by the application of sprays of water to the exterior of such vessel or vessels."

Administering a
lie detector test

Polygraph Lie Detector

Patent Name: Apparatus for Recording Arterial Blood Pressure
Patent Number: 1,788,434
Patent Date: January 13, 1931
Inventor: Leonard Keeler, of Berkeley, California

What It Does

The polygraph lie detector measures a variety of physiological fluctuations while an individual is undergoing questioning for investigative purposes, in order to help determine whether the person is answering the questions truthfully.

Background

The art of lie detection was introduced to science in 1921. A University of California medical student named John Larson invented the first polygraph lie detector—a machine that measured blood pressure, pulse, and respiration changes in response to questioning. The basis of the device is the belief that humans, when lying while under investigation, produce revealing involuntary physiological reactions that can be detected by experts. Leonard Keeler was a criminologist who ardently supported Larsen's invention and the theory behind it. He subsequently refined the invention, patented it, and marketed it commercially. Keeler's invention—initially used for medical and psychological testing only—was a primitive yet significant first step in the evolution of the polygraph.

While polygraph lie detectors are commonly used to interrogate crime suspects, they are not considered completely reliable. Some argue that the most criminal-minded individuals are free from relatively normal psychological constraints and are thereby able to control their physiological responses in order to "beat" the machine. On the other hand, individuals telling the truth under pressure of interrogation might show physiological signs that incorrectly implicate them. Lie detection remains an art, and the polygraph lie detector test is not generally admissible as evidence in court.

How It Works
1. A sphygmograph attaches to a person's body to measure blood pressure.
2. It is also connected to a recording device.
3. Fluctuations in pressure greater than those caused by the normal cardiac cycle are registered on paper by the recording device.

In the Inventor's Words
"It is an object of the invention to provide means whereby the sphygmogram or cardiac cycle may be recorded simultaneously with and be superimposed on the slower oscillations in the arterial pressure, whereby the characteristics of each as well as their relation to each other at any moment may be ascertained."

Post-it Notes

Patent Name: Acrylate Microsphere-Surfaced Sheet Material
Patent Number: 3,857,731
Patent Date: December 31, 1974
Inventors: Roger F. Merrill, Jr., of Troy Township, Wisconsin, and Henry R. Courtney, of St. Paul, Minnesota; assigned by Minnesota Mining and Manufacturing Company

What It Does
The Post-it note is one of the most quickly adopted and widely used office innovations of all time. Coming in different colors and sizes, the notes adhere to documents, computer monitors, doors, windows, and desk drawers throughout the world to help individuals on every rung of the corporate ladder. They are used to remind interns to send emails, CEOs to attend board meetings, and even husbands to buy milk.

Background   In 1970, research was being conducted at the Minnesota Mining and Manufacturing Company labs to improve the acrylate adhesives that the company used in many of its tapes. It was this company, 3M, after all, that first developed cellophane tape. Spencer Silver came up with an adhesive that didn't quite work—his stickum in the form of tiny spheres was just too weak to adhere well to any tape backing, because of the intermittent nature of the contact the spheres made with the backing. Despite the fact that his invention was considered too feeble to have any commercial value, Silver didn't give up on it. He lectured on its possibilities and praised its potential as a spray adhesive. His "acrylate copolymer microspheres" invention was awarded U.S. patent 3,691,140 in 1972.

Shortly thereafter, Art Fry, one of 3M's product development researchers, came to see the value in Silver's adhesive: It could be used to make a better bookmark. Corporate folklore holds that Fry was always frustrated by losing his place in the hymnal at church. From Fry's ideas, others flowed, and soon the Post-it note was born, hitting office-supply store shelves in 1980.

Arthur Fry, who helped develop the Post-It note

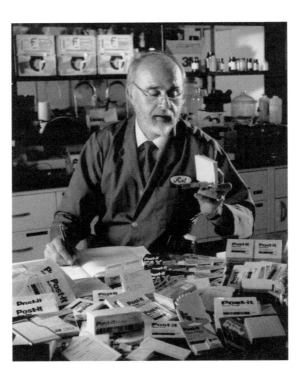

How It Works   1. Elastomeric copolymer spheres (30) range in diameter from 1 to 250 microns.
2. These are set upon a substrate (20), which adheres to a bind (30).
3. The substrate is not so porous as to allow the elastomeric copolymer spheres to permeate its surface completely.

Fashion designer Ilze Vitolina created a line of avant-garde evening wear fashioned from Post-it notes. To create the clothing, the notes were placed on a durable plastic, and plastic strips were used as fasteners. Vitolina created eleven dresses—including a wedding dress—as well as several hats and a bridal bouquet. The dresses were modeled at a 2000 fashion show sponsored by 3M Latvia.

4. Some substrates listed in the patent include films such as polyesters, cellulose acetate, and polyvinyl chloride, glass, wood, vinyl copolymers and urethane cast closed-cell foams, and paper.

In the Inventors' Words

"The invention relates to pressure-sensitive sheet constructions. More particularly, it relates to sheet constructions containing an adhesive system which permits repeated cycles in which materials are bonded thereto and removed therefrom.

"Conventional adhesives for adhering paper and other like materials to substrates, while having many desirable features, also have inherent drawbacks. For example, while some such adhesives may permit removal of paper from a substrate to which it has been adhered, they do not permit rebonding of the paper to the substrate. Conversely, other adhesives possess a tack which may be so aggressive as to cause the paper to tear on removal."

A remote
control store
demonstration
in 1960

Remote Control

Patent Name: Ultrasonic Transmitter
Patent Numbers: 2,821,954 and 2,821,955
Patent Date: February 4, 1958
Inventors: Robert Adler, of Northfield, Illinois; and Robert C. Ehlers, Lombard, Illinois, and Clarence W. Wandrey, of Wheaton, Illinois; assigned by Zenith Radio Corporation

What It Does   These two patents represent the innovation introduced as the "Zenith Space Command," in 1956—the first commercially successful wireless remote-control device for controlling a television signal from a short distance.

Background   Prior to the "Zenith Space Command," the company made the "Lazy Bones" remote, which was attached by a cable to the TV set. Then, Zenith introduced the "Flashmatic"—the first wireless remote, which used photocells and had many drawbacks. Mildly stated, the "Flashmatic" was erratic. Austrian-born Zenith research director Robert Adler championed the development of ultrasonic technology to perfect the remote. Ultrasonic waves—sounds that extend beyond the range of human hearing—provided the standard for two and a half decades of channel surfing until infrared light technology was introduced in the early 1980s.

Manipulating one's televised entertainment was not so easy before the remote control. Today, many people in privileged societies cannot imagine having to actually get up to change the channel, and legions of couch potatoes are probably unaware of the debt they owe Robert Adler. While Adler was awarded more than one hundred patents for a variety of technological innovations, he is regarded in the television industry as the father of the remote control.

Feb. 4, 1958   R. C. EHLERS ET AL   2,821,955

ULTRASONIC TRANSMITTER

Filed March 11, 1957

How It Works   Before transistors began to replace vacuum tubes, the tubes in the remote-activated TV sets were designed to receive and process the ultrasonic signals transmitted by the remote control.

1. The transmitter consists of four aluminum rods.
2. Each rod is carefully cut to lengths to generate a distinctive ultrasonic frequency.
3. Buttons on the remote-control unit engage a tiny spring-activated hammer that strikes the corresponding aluminum rods, thereby creating the high-frequency sounds.
4. Each of the rods has a distinct function: One turns the channel up; one turns it down; one increases volume; one decreases volume.

In the Inventor's Words   "I think a large part of the public said, 'An extra $100? For that, I can get up from my chair!' So [the first remote control] was never a large percentage of the sets sold." —Robert Adler

Self-Driving Car

Patent Name: Vision System for an Autonomous Vehicle
Patent Number: US 20070291130 A1
Patent Date: December 20, 2007
Inventors: Alberto Broggi, Gary Schmiedel, and Christopher K. Yakes, of Parma, Italy

What It Does

This system uses sensors and navigation software to allow an automobile to drive on roadways and through traffic without any direct human input.

Background

There are hundreds of patents for technologies used in autonomous cars, but the first company to put a self-driving car on public roads that navigated a mixed-traffic route was the Artificial Vision and Intelligent Systems Laboratory, or VisLab. The first U.S. patent critical to VisLab's autonomous car, called BRAiVE, was for a camera and sensor system that could take images of the surrounding area and convert that data into control input commands for a central vehicle management computer. In July 2013, BRAiVE autonomously drove through downtown Parma, Italy, navigating two-way streets, traffic lights, crosswalks, roundabouts, and other obstacles.

Many large automotive and tech companies have patents related to self-driving cars, such as Volkswagen's 2000 patent for "autonomous vehicle arrangement and method for controlling an autonomous vehicle." However, VisLab's computer-vision systems, or similar technologies, are central to all self-driving cars. The Italian company holds patents from U.S. and European patent offices, including a U.S. patent for "vehicle diagnostic based on information communicated between vehicles." The ability of autonomous cars to share data, as outlined in this patent, is considered crucial to deploying self-driving vehicle fleets across large areas such as entire cities.

Tesla, Inc., owned by inventor and engineer Elon Musk, has already started implementing self-driving technologies through its Autopilot feature, which will drive a passenger to a GPS location with little or no input from the driver. Tesla, however, has not patented any of its self-driving technologies, choosing rather to preserve research related to Autopilot within the company. Toyota has filed the most autonomous-car patents, with 54 issued out of 121 applications as of November 2016. Tech companies such as Google and Amazon have also filed patents for autonomous-car technologies, including a Google patent published in December 2016 for "determining pickup and destination locations for autonomous vehicles."

It is widely predicted that self-driving cars will change daily life more than any technological innovation since the original iPhone. Proponents of autonomous vehicles argue that a world filled with self-driving cars will significantly reduce automobile-related accidents and deaths. A network of self-driving cars that share navigation data has the potential to eliminate traffic in urban areas as well. While technologies like Tesla's Autopilot allow self-driving cars to navigate ordinary streets along with human drivers in conventional cars, many predict that future cities will eventually ban conventional cars in favor of a more seamless, efficient, and safe network of self-driving cars.

FIG. 9

FIG. 10

How It Works

1. A self-driving car incorporates three primary technologies: a GPS navigation system, a computer-vision system to identify the dynamic conditions of the road, and a central computer system to translate data into movement commands for the automobile.

2. The GPS system identifies a starting and ending destination and analyzes the various possible routes to select the preferred one.

3. Various technologies, including radar, cameras, and laser scanning, work in conjunction to compile information about the variables surrounding the car, such as traffic signals, pedestrian locations, lane spacing, moving obstacles, and road conditions.

4. The computer system, often called a controller area network (CAN) bus, runs complicated algorithms to translate the enormous amounts of raw data into movement commands, such as acceleration, braking, turning, and even parking.
5. Eventually, multiple autonomous cars will be able to share traffic and road data, to more efficiently navigate the roadways.

In the Inventor's Words

"We originally started working on driver assistance systems…and we developed a few applications like the lane detection, obstacle detection, and pedestrian detection thing, and then we moved toward the autonomous driving. We put everything together and tried to see if we could have some vehicle drive by itself using all these systems." —Alberto Broggi

Solar Panel

Patent Name: Apparatus for Utilizing Solar Radiant Energy
Patent Number: US 389124 A
Patent Date: September 4, 1888
Inventor: Edward Weston, of Newark, New Jersey

What It Does

This panel of photovoltaic cells absorbs photons of light from the sun and releases electrons, generating electricity.

Background

The history of solar panels dates back to nineteenth-century French physicist Edmond Becquerel's experiments. Becquerel discovered that certain materials would generate a small electric current when exposed to light, a phenomenon known as the photovoltaic effect. In 1839, Becquerel created the world's first photovoltaic cell—silver chloride placed in an acidic solution and connected to platinum electrodes.

The first U.S. patent for a solar cell was awarded to Edward Weston in 1888. It describes a "thermoelectric element wherein two bodies of dissimilar metal are placed side by side united at one end, and everywhere else insulated from one another" to "cause an electrical current in a circuit" when the device is exposed to sunlight. Weston even describes a battery-like storage system so that "energy accumulated during hours of sunshine may be utilized during night or periods of cloudy weather." The problem of storing solar energy for later use remains the biggest roadblock to widespread implementation of solar farms, even today.

Since the 1800s, significant improvements have been made to photovoltaic cells, primarily in the semiconductor materials used. Today's solar panels are made primarily of silicon, first in a crystalline form, and more recently in a thin film form that can be deposited in layers known as amorphous silicon. Solar cells gained widespread recognition after the Vanguard 1 satellite was launched in 1958—the first spacecraft to use solar power.

The largest solar power plant in the world, India's Kamuthi Solar Power Project, uses photovoltaic technology and silicon-based solar cells. The solar farm covers about 3.9 square miles and has a power capacity of almost 650 megawatts. China is currently expanding its Longyangxia Dam Solar Park, which is expected to become the largest in the world when completed, and countries all over the world continue to invest in solar power plants and infrastructure to replace fossil fuels.

How It Works

1. A thin film of amorphous silicon is deposited between two electrode sheets on a substrate.
2. The electrode sheets and silicon semiconductors are connected by additional electrodes that penetrate through the layers, creating a circuit.
3. The thin semiconductor film is treated to form an electric field, with one positive side and one negative side.
4. When sunlight hits the solar cell, electrons are released from the atoms in the silicon.
5. The released electrons are captured by the circuit as an electric current that can be used to power another device or can be stored in a battery for later use.
6. Multiple solar cells can be electrically connected and mounted to form a photovoltaic module, which can in turn be connected to form a solar array.

In the Inventor's Words

"I propose to transform radiant energy derived from the sun into electrical energy, or through electrical energy into mechanical energy. I may directly employ the electrical energy so obtained, or I may convert it into mechanical energy prior to such utilization."

SURGICAL DRESSINGS FOR WAR RELIEF

Tampons were used to apply antiseptics during surgery, as seen in the top right of this ad.

Tampon

Patent Name: Catamenial Device
Patent Number: 1,926,900
Patent Date: September 12, 1933
Inventor: Earle C. Haas, of Denver, Colorado

What It Does

Provides handy, hygienic, and undetectable internal means of absorbing menstrual flow.

Background

Some form of tampon has been in use for ages. While the men of ancient Egypt were covering the walls of pyramids and obelisks with hieroglyphics, their women were crafty enough to insert plugs of softened papyrus to absorb their menses. Women all over the world have used whatever natural material was handy for this purpose. External cloth pads eventually became the prevailing preference, and these "sanitary napkins" became commercially available in the 1920s—but they were neither practical nor comfortable.

At the same time, doctors regularly used makeshift plugs of cotton to apply antiseptics in bodily cavities, to stanch hemorrhaging, and to absorb blood during surgical procedures. It was in fact an observant physician, Dr. Earle Cleveland Haas, who put two and two together—his sympathy for his wife's dissatisfaction with her pads led him to invent the first disposable tampon with an applicator, very similar to those used today. Calling upon the Greek term for the monthly menstrual discharge, he called his invention a "catamenial" device, but the parlance changed when he registered a trademark for his invention. Combining the words "tampon" and "vaginal pack," Haas put Tampax on the market, where it has remained ever since.

How It Works

Toxic shock syndrome came to light in 1980, after the advent of a highly absorbent new tampon product called Rely. TSS is caused by a common bacteria—*Staphylococcus aureus*. Certain strains of this bacteria can produce toxins that poison the blood, causing nausea, fever, diarrhea, mental confusion, rash, and in some cases, death. It is believed that leaving a tampon in too long or using one with higher absorbency levels may allow the rare strain of bacteria to flourish. Rely was removed from shelves in 1981, but warnings are now printed on the packaging of all tampon products, advising frequent changing and the monitoring of possible symptoms. Dozens of women have died from toxic shock syndrome, but most women are willing to assume the risk in order to avail themselves of this comfortable and convenient product.

1. A seam (11) is sewn through an elongated strip of cotton (10).
2. An additional length of string (12) extends beyond the strip.
3. The strip is compressed under high pressure to form and retain a cylindrical core (13) until use.
4. The core is placed in an outer tube (14).
5. Another tube (15) is placed within the outer tube, protecting the sterile core.
6. A rounded edge (16) on the outer tube increases comfort upon insertion.
7. The inner tube operates as a plunger, facilitating insertion, after which both tubes are removed and disposed of.
8. For removal, the string (12) provides a means of drawing the tampon out.

Sept. 12, 1933.

E. C. HAAS

CATAMENIAL DEVICE

Filed Nov. 19, 1931

1,926,900

Inventor

EARLE C. HAAS

By

Attorney

In the Inventor's Words

"The principal object of this invention is to provide an absorbent pad and means for conveniently inserting said pad into the female vagina.

"Another object of the invention is to combine as a single unit an absorbent pad for insertion in the vagina with an inserting applicator, which acts as a container for the pad, so that the pad can be furnished in the applicator and inserted without removing it therefrom and so that the applicator can be discarded after use."

A traffic light on Fifth Avenue in New York City in 1933

Traffic Signal

Patent Name: Traffic Signal
Patent Number: 1,475,024
Patent Date: November 20, 1923
Inventor: Garrett A. Morgan, of Cleveland, Ohio

What It Does

Signals hung prominently at active intersections provide an orderly and safe traffic flow.

Background

Garrett Morgan also invented a gas mask that proved sensationally effective on July 15, 1916, when he used it to rescue people in an underground tunnel beneath Lake Erie after an explosion occurred.

Shortly after automobiles began rolling out of factories and onto public streets, traffic safety became a major issue. Until then, horse-drawn carriages and bicycles shared the roadway in relative harmony. A collision between an automobile and a carriage was the impetus for an African American named Garrett Augustus Morgan to invent the traffic signal.

The son of former slaves, Morgan was born in Kentucky and moved to Cincinnati and then to Cleveland while still a teenager. He worked as a sewing machine repairman, developing a reputation as a crafty and innovative mechanic. He ultimately became a successful businessman with a notable concern for public safety.

Though it wasn't the green, red, and yellow signal now used universally, Morgan's traffic signal did consist of three separate indicators: stop, go, and an all-direction stop to allow for street-crossing pedestrians. This last indicator also served the purpose of the current yellow light: a means to warn drivers that the flow of traffic was changing. A traffic conductor easily operated the signal by turning a crank.

How It Works

1. A rigid, boxlike structure is mounted atop a supporting pole at an active intersection.
2. Indicators feature the words "Stop," "Go," and "All Stop" on alternating panels.
3. The panels are supported by pivots operated by a hand crank.

4. The bottom of the supporting pole is affixed into a stand, out of which the crank handle protrudes. An operator uses the crank to raise the indicating arms, by means of ratchets; to lower them; and to revolve them.

5. The direction-indicating arms are pivotally supported and adapted to move vertically for stopping the flow of traffic in one direction; and then revolved and dropped to indicate a right-of-way to vehicles moving in another direction.

6. The box on the top is rotated by turning the crank, which engages a circular ratchet mechanism, so that the panels with the traffic instructions can face alternating different directions.

7. Traffic is thus directed by raising the arms to stop traffic in one direction, whereby all traffic is stopped; then revolving the arms, which also rotates the top box, and releasing them to indicate right of way for traffic in another direction.

In the Inventor's Words "One of the objects of my invention is the provision of a visible indicator which is useful in stopping traffic in all directions before the signal to proceed in any one direction is given. This is advantageous in that vehicles which are partly across the intersecting streets are given time to pass the vehicles which are waiting to travel in a transverse direction; thus avoiding accidents which frequently occur by reason of the over-anxiety of the waiting drivers to start as soon as the signal to proceed is given."

AFRICAN AMERICAN INVENTORS

The inventive contributions of African Americans are often overlooked. Here is a list of just a few important ones:

Thomas L. Jennings patented a process of dry-scouring clothing. With his earnings, he was able to extricate himself and his family from the bonds of slavery.

A nuclear physicist, **George E. Alcorn**, invented an imaging X-ray spectrometer.

Lewis Latimer

Thomas Edison may be famous for inventing the lightbulb, but without **Lewis Latimer** they would be more expensive, less efficient, and die more easily. He invented a process to make carbon filaments.

After quitting school at age ten, **Granville T. Woods** invented the incubator and made a career move from firefighter to engineer.

Hermon Grimes invented the folding wing aircraft, the design of which was used in a craft flown by George H. W. Bush during World War II.

Richard B. Spikes invented the automatic gearshift and a braking device still used in school buses.

Dr. Charles Richard Drew was the first African American surgeon to serve as an examiner on the American Board of Surgery. He invented the concept of blood banks.

Dr. Charles Richard Drew, the inventor of blood banks

Percy Julian was a chemist and inventor who used soy to make revolutionary products, such as aeroform, a flame retardant that saved lives during World War II, therapies for miscarriage and cancer, and soy-based cortisone treatment for arthritis.

Madam C. J. Walker—the daughter of two slaves—lost her parents at a young age and moved around the country suffering abuse and poverty. Due to all of this stress, she started losing her hair. She had a dream where she learned a formula that would make her hair grow back and it not only worked for her, but all of her friends. She realized that there were no beauty products for African American women, so she started the Madam CJ Walker Manufacturing Company.

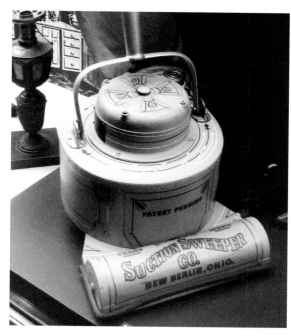

An early vacuum, called the Electric Suction Sweeper, from 1908

Vacuum Cleaner

Patent Name: Carpet Sweeper and Cleaner
Patent Number: 889,823
Patent Date: June 2, 1908
Inventor: James M. Spangler, of Canton, Ohio

What It Does

Sucking dirt, dust, and debris out of the carpets and furnishings of homes, hotels, and offices around the globe, the vacuum cleaner is one of the handiest sanitation mechanisms ever invented.

Background

An asthmatic department store janitor named James Spangler invented the modern electric vacuum cleaner in 1907, and established the Electric Suction Sweeper Company. The president of the company was an enthusiastic customer of Spangler's product and also his cousin's husband—a man named William Hoover. Hoover is credited with the concept of door-to-door salesmanship. To attract attention to the product, Hoover offered in-home demonstrations. In 1922, he changed the company name to "Hoover." Spangler's name, on the other hand, got sucked out of the spotlight.

How It Works

1. An electric motor (2) attaches to a fan casing (1).
2. From the motor, a shaft (3) protrudes into the casing, where it attaches to a hub (3).
3. The hub is securely connected to the fan blades (5), which are arranged, when rotating, to create suction at the casing's bottom.
4. The casing sits on back wheels (34) and a gearwheel cylinder covered in a circular brush (12).

5. Also at the bottom, a plate (13) is arranged to funnel debris into the machine through an inlet opening (14).

6. Dust and debris are then blown through an outlet (28, in figure 1) and into a collecting sack (27).

In the Inventor's Words

"In a sweeper, the combination with a casing formed with a centrally disposed opening in its bottom, a bearing extending over the opening, a vertical shaft journaled into the bearing, a fan mounted on the vertical shaft, a flat housing which tapers toward the front and located under the casing which communicates with the centrally disposed opening, said housing extending beyond the casing and formed with an elongated opening in its bottom at the widest portion, a brush mounted in the extended end of the housing adjacent the elongated opening, means of communicating with the casing for receiving the dust laden air, and means for revolving the brush."

Viagra

Patent Name: Pyrazolopyrimidinodes Which Inhibit Type 5 Cyclic Guanosine 3',5'—Monophosphate Phosphodiesterase (cGMP PDE5) for the Treatment of Sexual Dysfunction
Patent Number: World Intellectual Property Organization 98/49166-A1
Patent Date: November 5, 1998
Inventors: Mark Edward Bunnage, John Paul Mathias, Stephen Derek Albert Street, and Anthony Wood, all of Sandwich, Kent, United Kingdom: All patentees assigned by Pfizer Central Research

What It Does

Viagra and other similar drugs are becoming an increasingly popular treatment for erectile dysfunction. The drug essentially enables a fuller flow of blood to the penis, promoting more reliable and longer-lasting erections.

Background

In 2003, Wrigley's Gum was granted a U.S. patent for a gum that delivers sildenafil citrate. Who knows? ED sufferers may soon be chewing their way to better sex lives.

Blue versus Yellow In 2003, the NFL became the first professional sports league to sign an endorsement deal allowing GlaxoSmithKline and Bayer Pharmaceuticals Corporation to market a Viagra rival called Levitra. Levitra, which comes in the form of a little yellow pill, competes with Pfizer's little blue pill.

When Masters and Johnson released their revolutionary studies of sexuality in the 1950s, private topics suddenly became the subject of public discourse. This helped open wide the door to medical studies that focused on sexual behavior in general and reproductive organs specifically. Not surprisingly, commercial markets sprang up around products that held the promise of enhanced sexual satisfaction.

The brainchild of researchers at the pharmaceutical giant Pfizer, Viagra is one such product for which the company has fought long and hard. A medicine used to treat erectile dysfunction (ED), Viagra first became available in 1998. In the year 2012 alone, more than 8 million prescriptions were written for the little blue pill, generating $1.93 billion in revenue for Pfizer.

Clearly, Viagra is a moneymaker and the manufacturer has fought to monopolize the market for it. The United Kingdom's High Court ruled that Pfizer's European patent (involving the use of sildenafil, Viagra's active ingredient)

Female sexual dysfunction (FSD) is another well-documented problem, and Viagra has been helping women, too. New products are under development to address specifically female problems.

was invalid. The legal challenge was brought by Eli Lilly and ICOS Corp., who were codeveloping Cialis, a rival impotency drug. Other rival medications are in the works, and some are on the shelves already. The battle lines have been drawn and the swords are raised. In October 2003, Pfizer filed a lawsuit against GlaxoSmithKline and Bayer Pharmaceuticals Corporation for patent infringement, for the marketing of Levitra.

How It Works

Sildenafil, the active ingredient in Viagra, acts like a traffic director who shrewdly opens lanes in one direction during rush hour. The compound enables arteries in the penis to relax and widen to offer more access for blood flow toward the organ. Simultaneously, the drug restricts the veins that carry blood away from the penis. Altogether, sildenafil is known to achieve solid results during sexual intimacy.

In the Inventors' Words

"Compounds of formulae (IA) and (IB) or pharmaceutically or veterinarily acceptable salts thereof, or pharmaceutically or veterinarily acceptable solvates of either entity, wherein R_1 is C1 to C3 alkyl substituted with C3 to C6 cycloalkyl, $CONR_5R_6$ or a N-linked heterocyclic group; $(CH_2)nHet$ or $(CH_2)nAr$; R_2 is C1 to C6 alkyl; R_3 is C1 to C6 alkyl optionally substituted with C1 to C4 alkoxy; R_4 is $SO_2NR_7R_8$; R_5 and R_6 are each independently selected from H and C1 to C4 alkyl optionally substituted with C1 to C4 alkoxy, or, together with the nitrogen atom to which they are attached, form a 5- or 6-membered heterocyclic group; R_7 and R_8, together with the nitrogen atom to which they are attached, form a 4-R_{10}-piperazinyl group; R_{10} is H or C1 to C4 alkyl optionally substituted with OH, C1 to C4 alkoxy or $CONH_2$; Het is an optionally substituted C-linked 5- or 6-membered heterocyclic group; Ar is optionally substituted phenyl; and n is 0 or 1; are potent and selective cGMP PDE5 inhibitors useful in the treatment of, inter alia, male erectile dysfunction and female sexual dysfunction."

Six

The Patent That Pleases

A statue of James Naismith, the inventor of the game of basketball

Basketball

Patent Name: Basketball
Patent Number: 1,718,305
Patent Date: June 25, 1929
Inventor: George L. Pierce, of Brooklyn, New York

What It Does

This sturdy, well-designed, uniformly manufactured, and long-lasting basketball is the prototypical centerpiece to the modern game of basketball, both professional and amateur.

Background

NBA standards call for baskets that are 18 inches in diameter with the rim standing 10 feet above the floor. The ball itself must contain between 7.5 and 8.5 pounds of pressure.

The modern game of basketball was invented by Canadian James Naismith, a graduate of McGill University in Montreal. He took a job as a physical education instructor at the School for Christian Workers in Springfield, Massachusetts. A strict boss gave him orders and a short deadline to come up with a way to occupy a class of rowdy kids during the fast-approaching winter of 1891–92. Naismith came up with a set of rules and a game that is the closest known precursor to the modern sport of basketball.

As important as the rules, of course, were the tools: a smooth surface, even hoops, sturdy backboards, and a decent ball. As is often the case when new recreational sports receive considerable attention, companies began inventing and manufacturing products to meet the growing demand. One invention that has long pleased basketball players and fans is the improved seam-construction design in this patent awarded to George Pierce, who was acting as assignor to A. G. Spalding & Brothers. Prior to Pierce's unique stitching arrangement, the panels of hide on a basketball were tapered to polar points, which resulted in many inferior characteristics, such as the placement of a valve on the equatorial portion of the ball, and an altogether weaker structure. Pierce improved the design immeasurably by reshaping the panels.

How It Works

1. Panels are cut with abrupt, concavely curved ends.
2. The same panels include correspondingly curved projections at the centers.
3. Polar projections provide for a closure seam at one pole and an inflating valve at the other.

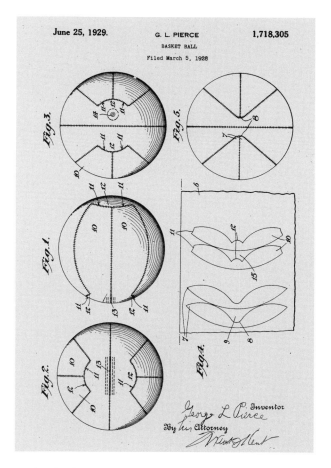

June 25, 1929.   G. L. PIERCE   1,718,305

BASKET BALL

Filed March 5, 1928

In the Inventor's Words

"What is claimed is: 1. A case for a game ball, composed of panels, each of which has length equal to a major part of the ball circumference, one continuously curved side edge, opposite, shorter, separately curved side edges, a convexly curved polar projection intermediate to the separately curved side edges, and concavely curved ends to fit portions of the polar projections of another panel, the panels being connected together in pairs along said continuously curved side edges, and separately curved side edges of the panels of one pair being connected to similar edges of panels of another pair, and the concavely curved ends of the panels of one pair being connected to the polar projections of panels of another pair…"

"Best to you each morning."

Cornflakes

Patent Name: Cereal Food and Process of Production
Patent Number: 1,321,754
Patent Date: November 11, 1919
Inventor: John Kellogg, of Battle Creek, Michigan

What It Does

Cornflakes are one of the world's most beloved breakfast cereals. Ingredients include milled corn, flavoring, iron, vitamin C, vitamin B_6, vitamin B_2, folic acid, vitamin A palmitate, vitamin B_{12}, and vitamin D, all mixed, toasted, and flaked to make a reassuringly bland but filling morning meal (much enhanced nutritionally with the addition of milk).

Background

Today, the company projects more than $13 billion in annual sales. It makes its products in nineteen countries, sells them in more than 150, and still exhibits some of its early vegetarian roots. Brand names the company markets include Kellogg's, Keebler, Pop-Tarts, Eggo, Cheez-It, Nutri-Grain, Rice Krispies, Special K, Kashi, and Morningstar Farms, which makes a variety of meat alternatives.

The story of cornflakes begins with two brothers and the mandates of the Seventh-Day Adventist Church at the turn of the twentieth century. Dr. John Harvey Kellogg administered a health reform sanitarium founded by his church in Battle Creek, Michigan. John was a stern and educated doctor, obstinately devoted to the tenets of his religion, who ruled his patients, staff, and brother with an iron fist. Less educated and less interested in fire and brimstone, his brother Will was a general maintenance worker and handyman around the place.

Will also became the cook, and his breakfast of choice consisted of hot cereals of grain, oat, or corn. In 1895, the morning after Will accidentally left some cereal on a hot stove, brother John discovered the flaky substance that

John Harvey Kellogg

resulted. He tasted it, deemed it palatable, and refined it, and the first flaked breakfast cereal was born. Cold cereal was a new thing back then, but the doctor blessed it and served it to the guests at the sanitarium. He believed the cereal would minimize his patients' desires to masturbate, a punishable no-no at the sanitarium.

As grain foods started turning profits, Dr. Kellogg attracted competition. His neighbor, C. W. Post, joined in the frenzy of moral cleansing by marketing Postum and Grape Nuts out of his own sanitarium. Meanwhile, Dr. Kellogg's own grain food company was losing money and the sanitarium was registering financial losses as well. In 1905, Will asserted his own business sense and convinced his brother to establish a new company to make cornflakes. The company was incorporated in 1906 and placed under Will's management. When the doctor was off on a trip, Will aggressively purchased stock from his fellow stockholder, making himself company president and putting his signature on the cereal box that now memorializes him.

How It Works

1. A quantity of grain is cooked long enough so that its starches become dextrinized.
2. One method of cooking commonly used is a steaming conveyor on which the grain is propelled through pressurized steam.
3. After it has sufficiently cooked and partially dried, flavoring ingredients may be added.
4. The cereal is then toasted to give the flake its crispiness.

In the Inventor's Words

"My invention relates to the production of a ready-to-eat cereal food from any suitable grain, such as corn, rice, oats, wheat, or barley.

"My invention consists briefly of a cereal food made and the process of making the same in the form of separate shreds, flakes, or particles of other form…"

The first drive-in movie theater, in Camden, New Jersey

Drive-In Movie Theater

Patent Name: Drive-In Theater
Patent Number: 1,909,537
Patent Date: May 16, 1933
Inventor: Richard Hollingshead, Jr., of Riverton, New Jersey

What It Does

Provides a novel means of outdoor entertainment by projecting a movie on a large open-air screen accompanied by personal listening devices so that people in parked cars can both see and hear the film.

Background

The second drive-in movie theater, the Shankweiler's Drive-In Theater in Orefield, Pennsylvania, was opened the year after Hollingshead's, in 1934. It is now the oldest continuously operating drive-in.

The drive-in theater with the largest capacity was the All-Weather Drive-In of Copiague, New York, a 28-acre lot that could accommodate 2,500 parked cars. It included a monumental indoor viewing area that could seat 1,200.

As a young man selling auto parts at his father's shop, Richard Hollingshead began experimenting with a little idea he had. He aimed a 1928 Kodak projector at a sheet stretched between two trees, and positioned a radio behind the sheet for sound. He even raised the front ends of cars on blocks for better viewing. Eventually, Hollingshead honed his ideas into a vision so precise as to include a method of keeping insects away from the film projector.

He broadly described his invention as "a novel construction to outdoor theaters whereby the transportation facilities to and from the theater are made to constitute an element of the seating facilities of the theater."

In other words, the audience is seated within the very means that brought them!

This aspect of Americana—the drive-in movie theater—combines our love affair with the automobile and our insatiable demand for entertainment. It has also helped jump-start many off-screen adolescent romances.

On June 6, 1933, less than a month after he was issued his patent, Hollingshead opened the first drive-in movie theater, in Camden, New Jersey, with a showing of *Wife, Beware*. The space could accommodate four hundred cars. Admission was twenty-five cents per person plus an additional twenty-five cents per auto—at this rate, a date could cost as little as seventy-five cents. Subsequent innovations, including in-car sound systems, increased the popularity of the drive-in theater; by the year 1958, just twenty-five years after the inventor opened the first one, more than 4,000 drive-in theaters were scattered across the United States and Canada.

By the 1980s, however, the lure of the drive-in theater was overshadowed by the convenience of home entertainment systems and the development of large indoor multiplex theaters. Urbanization made the land that the drive-in theaters occupied more valuable than the service they provided; owners were pressured to sell to developers. By 1990, only around nine hundred drive-in theaters remained. While there is some interest in reviving them, drive-in movie theaters have become primarily the stuff of nostalgia, cultural symbols of a bygone era.

How It Works

1. After dark, customers drive into prearranged slots in a large outdoor lot, slightly inclined and facing a screen.
2. A movie projector projects images onto a screen.
3. A central amplifier powers individual speakers temporarily attached to each car or, alternatively, centralized speakers project over a broad range.
4. Customers watch and listen from inside their cars.

The inventor included twenty specific claims in his patent application, nineteen of which were amended variations on the first: "An outdoor theater comprising a stage, alternate rows of automobile drive-ways, and vertically inclined automobile stall-ways arranged in front of the stage, said stall-ways being adapted to receive automobiles disposed adjacent to each other and facing the stage;—said automobile stall-ways being at a vertical angle with respect to the stage such as will produce a clear angle of vision from the seat of the automobile, through the windshield thereof to the stage, free of obstruction from the automobiles ahead of it."

Etch A Sketch

Patent Name: Tracing Device
Patent Number: 3,055,113
Patent Date: July 23, 1959
Inventor: Arthur Grandjean, of Paris, France

What It Does

This educational toy provides hours of amusement as children manipulate the dials to create erasable line drawings.

Background

Almost sixty years after its introduction in 1960, Etch A Sketch is still a popular product. The Canadian company Spin Master now produces and markets it.

Imagine you are looking through a closed window that is covered in dust. You'd like to put your finger to the dust and scribble a pattern into it, but the dust is on the opposite side of the glass. By design, the inventive principles employed by the Etch A Sketch would allow you to draw such a pattern from your side of the glass. Arthur Grandjean invented his L'Écran Magique (Magic Screen) in his garage in Paris, and showcased it at the 1959 International Toy Fair in Nuremberg, Germany. Two Americans representing the Ohio Art Company bought the rights to develop the invention, and the small toy company turned it into the now-classic silver-screened toy framed in red plastic with two white dials.

How It Works An adherent material, such as aluminum powder, sticks to the underside of the screen's surface. Under the screen is a stylus that can be moved by two knobs at the bottom of the frame to move the powder aside and create patterns on the screen from below. One knob controls vertical movement of the stylus; the other, horizontal. Turning the knobs simultaneously can produce diagonal lines and requires a degree of skill and coordination to accomplish a desired affect. To erase, simply shake the entire screen and the aluminum powder flows back into place.

1. A moveable stylus mechanism (10), in patent illustrations, directly contacts the surface of the glass.
2. The stylus can be moved along two rods (11, 12), which are arranged at right angles to each other and are adapted to move in orthogonal directions by means of two cable systems (13, 14).
3. Turning the knobs (17, 18) engages movement of the cables, which in turn engages movement of the rods to which the stylus is affixed.
4. The stylus scratches a line of aluminum dust from the screen, creating a visible line.

In the Inventor's Words "This arrangement, which is capable of many applications for advertising purposes, for moving pictures, toys, testing means for instance, is constituted chiefly by means of a fluid-tight case provided in its upper part with a translucent surface of glass or the like material which is made opaque when the case is turned upside down, said case being partly filled with a pulverulent metallic or other material adapted to adhere then to said translucent surface and being provided inwardly with a removable tracing stylus which engages frictionally the inner surface of the translucent surface, so as to remove through its passage the pulverulent material adhering thereto and to form lines which are visible from the outside of the case and may be immediately wiped out through a mere turning of the case upside down associated with a shaking of the latter."

Jimi Hendrix playing a
Fender Stratocaster
guitar

Fender Stratocaster

Patent Name: Guitar
Patent Number: D169,062
Patent Date: March 24, 1953
Inventor: Clarence L. Fender, of Fullerton, California

What It Does

This superior design for an electric guitar features an ergonomic and attractive body, and is widely considered to be an electric guitar of legendary quality.

Background

Repeatedly acquiring new Stratocasters, Jimi Hendrix must have been one of Leo Fender's most devoted customers. The left-handed musician was known to burn and smash his guitars during his electrifying performances.

Other Strat-playing stars include:

• Dick Dale
• Buddy Holly
• Stevie Ray Vaughn
• Eric Clapton
• Chrissie Hynde of the Pretenders

Unlike acoustic guitars, which depend on hollow-body resonance, electric guitars have solid bodies and depend on electromagnetic pickups and amplifiers to produce their unique and powerful sound. Though not a player himself, Clarence "Leo" Fender was an electric guitar aficionado who possessed a keen understanding of how the instrument worked. With Old World attention to craft, Fender made his guitars into masterpieces. For his New World business sense, he was sometimes called the Henry Ford of the electric guitar—he regularly updated his designs for the needs of an individual musician. His models were based on a variety of different guitar designs with funky names like the Broadcaster, the Telecaster and, most famously, the Stratocaster.

Featuring an attractive and ergonomic design, the Stratocaster came to include components that rival electric guitars lacked. A bolt-on neck allows for speedier manufacture and if the neck happens to warp, the entire instrument need not be trashed. Three electromagnetic

No design changes were made to the Stratocaster until the 1980s, and even then, the changes were quite subtle.

pickups are strategically placed to transfer the vibrations at bass, midline, and lead positions into electric signals. Covers for the pickups help eliminate undesired feedback, and higher notes become suddenly more accessible. The Fender Stratocaster is considered to make a unique sound that helped launch the psychedelic and surf rock genres. Indeed, the Stratocaster was the instrument of choice for electric surf guitar guru Dick Dale and the left-handed rock virtuoso Jimi Hendrix.

How It Works

1. The body is usually constructed of a hard wood, such as maple.
2. Bright car paint may be applied to pleasing effect.
3. Electric components are added to the construction.
4. Strings are attached and tuned.
5. The body plugs into an amplifier.
6. The musician plays an altogether superior guitar.

In the Inventor's Words

"Be it known that I, Clarence L. Fender, a citizen of the United States, residing at Fullerton, in the county of Orange and State of California, have invented a new, original, and ornamental Design for a Guitar, of which the following is a specification, reference being had to the accompanying drawing forming a part thereof.

Figure 1 is a top plan view of a guitar embodying my new design;
Figure 2 is a side elevation view thereof;
Figure 3 is a bottom view thereof; and
Figure 4 is an end elevation view thereof.
I claim: The ornamental design for a guitar, as shown."

In 1890, children listen to an early jukebox in Salina, Kansas.

Jukebox

Patent Name: Coin-Actuated Attachment for Phonographs
Patent Number: 428,750
Patent Date: May 27, 1890
Inventors: Louis Glass and William S. Arnold, of San Francisco, California

What It Does

The jukebox is a coin-operated music machine, usually found in restaurants or other public establishments, which allows a customer to select from an inventory of single songs and play them through public speakers.

Background

Louis Glass and William S. Arnold pioneered a cultural phenomenon when they installed and demonstrated a coin-operated phonograph in San Francisco's Palais Royale Saloon on November 23, 1889. It was a far, and quiet, cry from what we call a jukebox today: The nickel-in-the-slot machine contained only one selection that a maximum of four people were able to listen to simultaneously through "hearing tubes." Amplification had yet to be invented.

The days of hip-shaking juke joints were still far off. The machine appeared decades before radio had hit the scene and just twelve years after Edison's phonograph was invented. It would be a couple of decades before the phonograph became a regular feature in the household. Its timing made the nickel-in-the-slot novelty a huge hit. Patrons enjoyed music, proprietors attracted extra revenue, and a nascent music-recording industry inadvertently found one of the earliest ways to market popular songs.

As amplification did away with the earphones, and song capacity increased, chic designs came to give the jukebox a sense of style. Part record player, part vending machine, part furniture, the jukebox established itself as a valuable and attractive component of nightlife culture. While the jukebox has gone through countless mutations, the concept of dropping a coin in a slot to hear a good tune in public has remained perennially popular.

How It Works

Time magazine first printed the term "Juke Box" in 1939—an adaptation of the southern colloquialism "jook," which means "to dance."

Early jukeboxes spun platters, though today's models feature CDs.

1. A mechanism for rendering the phonograph alternately operative and inoperative consists of a sliding piece that compresses the listening tubes.
2. A rock-shaft consists of two arms, one engaging pressure on the sliding piece, one projecting into the coin chute.
3. The coin is dropped into the chute, striking one arm of the rockshaft.
4. The other arm releases pressure from the sliding piece, allowing a recording to be heard through the listening tubes.

In the Inventors' Words

"Our invention relates, generally, to the class of devices designed to be operated by a suitable coin deposited properly, and especially to an attachment of this class intended to be operated in connection with a phonograph.

"Our invention consists in the novel constructions and combinations hereinafter fully described, and specifically pointed out in the claims.

"The object of our invention is to provide a suitable device by which the phonograph may be exhibited and heard by any one upon the deposit of a suitable coin."

Peach

Patent Name: Peach
Patent Number: PP15
Patent Date: April 5, 1932
Inventor: Luther Burbank, California

What It Does
A member of the Rosaceae family (to which the rose belongs), the edible delicacy known as the peach is as sweet as its flowered cousin is pretty. The fruit was first cultivated in China.

Background
Who doesn't love a peach? Or a plum? Or a potato, for that matter? For these and other healthy foods, we owe a great debt to Luther Burbank, born in 1849. Obviously, Burbank did not "invent" the fruits of Mother Nature, but he did pioneer the field of horticulture—the science of cultivating plant life—to bring many edible varieties and create viable year-round markets for farmers everywhere. He experimented by grafting the seedlings of various plants onto other fully developed plants, often obtaining superior hybrids. More than eight hundred plant strains and varieties were created by this singular man.

Helping Ireland to recover from the black rot epidemic that wiped out its potato crops and caused massive starvation, Burbank developed a new strain of potato, selling the seeds for this forerunner to the Idaho potato for a meager fee. His earnings allowed him to move from Massachusetts to California, where he continued to work in horticulture—but he received no patents in his lifetime, as plant strains were not yet patentable.

After creating many new strains of fruits, vegetables, and other plants, Burbank died in 1926, a mere four years before the Plant Patent Act was instituted. Burbank received several patents, including this one, posthumously. The act, meant to provide financial incentives to farmers and horticulturists, has itself borne much fruit, as Thomas Edison predicted in supportive testimony for the new legislation: "This [bill] will, I feel sure, give us many Burbanks.'

How It Works

Burbank's strain of peach is grown from a tree that is larger than other peach tree varieties of the same age. Its branches are stout and its twigs contain medium-to-large dormant fruit buds before bearing the fruit.

In the Inventor's Words

Luther Burbank was the father of horticulture, growing superior hybrids of produce.

"This new variety of peach has resulted from years of experimenting with a definite objective in mind, that is, to produce a satisfactory yellow freestone peach which ripens halfway between the ripening periods of the known varieties, the June Elberta and the Early Elberta. It is similar to the Hale peach except that it has a large pit. It blood and seed are similar to the Muir, but the fruit is more golden in color. It is a stronger growing tree than the Valient and is not subject to peach curl and disease (Bacteria impruni) as is the last named variety. This new variety produces a very large fruit which averages about one-half pound. Its golden color with maroon shadings modified by a grayish pubescence, adds to its effectiveness in size. Although the skin of the fruit is thin and tender, tests have proven it to be a remarkable shipper; coupled with its great size, impressive coloring, excellent quality, and being a freestone, it represents an outstanding commercial peach. When cut in half, a pleasing modified apricot yellow flesh is disclosed which has a peach red tinge near the pit."

Edison and the phonograph in 1878

Phonograph

Patent Name: Improvement in Phonograph or Speaking Machine
Patent Number: 200,521
Patent Date: February 19, 1878
Inventor: Thomas A. Edison, of Menlo Park, New Jersey

What It Does

The phonograph was the first machine to record and play back sound, using principles similar to those of the telephone, along with an amplifying mechanism. It was eventually modified only to play sound recorded elsewhere professionally onto vinyl disks.

Background

A machine that can reproduce sound when a stylus is applied to the grooves in a disk, the phonograph stands among Edison's greatest inventions. Its introduction sent Edison riding ever higher on the waves of international fame while it single-handedly spawned the birth of the recording industry. Not only does his patent describe the predecessor to the vinyl LP and record player, it describes similar features of a tape recorder. The invention could record and transcribe sound. It began as an offshoot of some experiments Edison was conducting in 1877 to improve his telephone and the diaphragm it used to transmit the sound of the voice. One inventive instinct led to another, and one can only wonder at the joy and satisfaction Edison must have felt when his stylus and tinfoil cylinder reproduced his own recorded voice intoning "Mary had a little lamb."

How It Works

To record:

1. A cylinder (a) has a helical indenting groove cut from end to end. (Edison offered an estimate of ten grooves per inch.)

Established in 2002 by a nonprofit organization of the same name, Vinyl Record Day is celebrated on August 12, what is believed to be the date in 1877 when Edison successfully invented the phonograph. In 2003, the Internal Revenue Service called the California-based nonprofit group the first vinyl-related tax-deductible organization.

Many people still prefer vinyl records to even MP3s, claiming that vinyl offers a more authentic sound quality.

2. The material (tinfoil in Edison's invention) to be indented is placed over the cylinder and secured to a shaft (X).
3. At one end, the shaft has a thread cut into it at about ten threads per inch.
4. A pillar bearing (P) also has a single thread cut into it.
5. A tube (L) can be fitted over the shaft to be rotated by means of clockwork or another power source (M).
6. A speaking tube (B) contains a diaphragm with an indenting point secured at its center.
7. The material passes between the pillar bearing and the indenting point as the machine is powered, registering the vibrations being indented upon it.

To play:

1. When the sound has been imprinted, it can be reproduced through the listening tube (C).
2. Like the speaking tube, the listening tube contains a diaphragm, but instead of an indenting point, a light spring with a stout point is attached at its center.
3. When the cylinder rotates, the spring is set in motion to correspond with the indentions and thereby produce a facsimile of sound.

To mass-produce:

1. The recording may be stereotyped by plaster-of-paris molds.
2. New sheets of foil can be pressed from these "master" molds.

In the Inventor's Words "The invention consists in arranging a plate, diaphragm, or other flexible body capable of being vibrated by the human voice or other sounds, in conjunction with a material capable of registering the movements of such vibrating body by embossing or indenting or altering such material, in such a manner that such register-marks will be sufficient to cause a second vibrating plate or body to be set in motion by them, and thus reproduce the motions of the first vibrating body."

Quadcopter Drone

Patent Name: Omni-Directional, Vertical-Lift, Helicopter Drone
Patent Number: US 3053480 A
Patent Date: September 11, 1962
Inventor: Edward G. Vanderlip, of Radnor, Pennsylvania

What It Does This small remotely piloted aircraft with four propellers is flown for recreational or commercial purposes.

Background Unmanned aerial vehicles (UAVs) have taken many shapes and forms, and the majority of remotely piloted aircraft were developed by the U.S. military. However, the military has primarily made use of fixed-winged drones, while most civilian and commercial drones are so-called "quadcopters." These small helicopter-type aircraft have four propellers spread around the main body to achieve lift. Some models use the same basic design with more propellers, but four is most common, and the earliest patent for what we would recognize as a common civilian drone had four propellers as well.

Piasecki Aircraft Corporation, the company where inventor Edward G. Vanderlip worked, built a number of conventional helicopters before developing Vanderlip's "omni-directional, vertical-lift, helicopter drone" in 1962. Vanderlip had received another patent, nine years earlier, for "automatic control for helicopters," which covered components designed to allow aircraft instruments and control systems to continue functioning properly in the event of engine failure. (When a helicopter engine fails, the pilot can still land using a technique called autorotation.) The first quadcopter drone was developed with a similar purpose, albeit for a remote pilot: to make flying easier.

The "helicopter drone having four lift rotors arranged in pairs at opposite ends of horizontal axes of tilt" was designed to be "extremely simple" to fly. The quadcopter drone achieves this by keeping its vertical axis pointed straight up while tilting its four rotors to fly in any direction without changing the aircraft's orientation. This configuration allows turns in flight without banking, and the drone has a constantly level platform for its electronics and control systems (something that has come in handy for the cameras attached to many drones today). In fact, quadcopters have become so simple to fly that children can easily learn to pilot them with a remote control, and civilian use has become so widespread that the U.S. Department of Transportation and the Federal Aviation Administration had to create a new set of regulations for small unmanned aircraft systems.

How It Works

1. A small remotely controlled aircraft uses four rotors to achieve lift and fly using vertical takeoff and landing (VTOL) capabilities.
2. The aircraft is designed to constantly keep the vertical axis aligned directly upward, to the sky.
3. The rotors tilt on this axis to allow the aircraft to fly in any direction without changing its orientation, making it very easy to turn and control the drone.
4. Various types of equipment and flight systems can be incorporated, such as cameras, motion sensors, and autonomous flight controls.

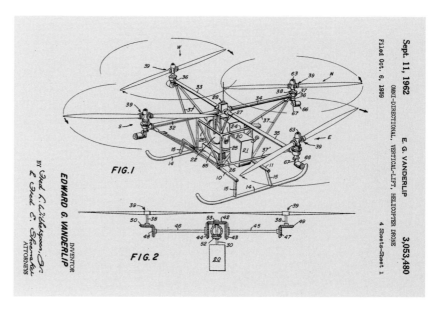

In the Inventor's Words

"The invention relates to the construction of and controls for a helicopter drone whereby it is caused to fly with its vertical axis at all times pointed to the zenith and with its orientation constant in spite of direction of flight, under which circumstances flight control is extremely simple, turns are made without banking, and a constant, level platform is provided for instruments, mechanical and electrical equipment, and the like, and all controls and manipulative equipment greatly simplified."

Radio

Patent Name: System of Signaling
Patent Number: 725,605
Patent Date: April 14, 1903
Inventor: Nikola Tesla, of New York, New York

What It Does   The radio broadcasts distinct electrical impulses through radio waves that carry sound over distances and can be heard when tuned at certain frequencies.

Background   Nikola Tesla rivaled his former employer Thomas Edison in a variety of electricity-based innovations; using a motor he invented and the energy of Niagara Falls, he successfully demonstrated hydroelectricity; and he invented a high-frequency "Tesla coil," still used today in radio and television transmission equipment. Yet, while Tesla was a brilliant inventor, he was never wealthy.

All too often, he was robbed of the credit he deserved for his inventions. The radio is one such example, demonstrating that Edison would not be Tesla's only rival. In this case, Italian inventor Guglielmo Marconi demonstrated a radio voice broadcast in San Francisco in 1895; newspaper reports hailed the phenomenon as a new invention. Never mind the fact that Tesla had demonstrated radio communication in Philadelphia two years earlier. Today, Marconi is still often recognized as the inventor of the radio. But Tesla's work speaks for itself: The radio, and the principles behind so many other technologies we employ today, would not be possible without Tesla's driving intellect.

With or without credit, Tesla continued experimenting and improving earlier inventions. The underlying principles of an earlier patent Tesla received for a system of transmitting electrical energy are the focus of this patent.

How It Works

1. In figure 1, two spirally wound conducting coils (S^1, S^2) connected with their inner ends to elevated terminals (D^1, D^2).
2. Their outer ends connect to an earth plate (E).
3. The two systems may have electrical oscillations impressed on them by energizing them through primaries (P^1, P^2) and regulated by inductances (L^1, L^2).
4. The two independent primary circuits send vibrations to the earth plate and spread to a distance reaching a similar circuit-forming receiving station tuned in conjunction with the sending station.

No. 725,605.

PATENTED APR. 14, 1903.

N. TESLA.
SYSTEM OF SIGNALING.
APPLICATION FILED JULY 16, 1900.

NO MODEL.

Witnesses:

Nikola Tesla, Inventor

by Ken. Page + Cooper
Att'ys

In the Inventor's Words

"Broadly stated, this invention consists in the combination of means for generating and transmitting two or more kinds or classes of disturbances or impulses of distinctive character with respect to their effect upon a receiving-circuit and a distant receiver which comprises two or more circuits of different electrical character or severally tuned, so as to be responsive to the different kinds or classes of impulses and which is dependent for operation upon the conjoint or resultant action of the two or more circuits or the several instrumentalities controlled or operated thereby."

INVENTOR PROFILE

**Nikola Tesla
(1856–1943)**

Nikola Tesla

Nikola Tesla was born in Croatia to a Serbian Orthodox priest father and an uneducated but highly intelligent mother. Tesla's greatest contribution to the field of electrical engineering was the discovery of the rotating magnetic field, which forms the basis of almost all alternating-current machinery.

Tesla was educated in Austria and at the University of Prague. While studying the Gramme dynamo, a direct-current generator that becomes an electric motor when reversed, he got the idea to begin experimenting with alternating currents. In 1882, Tesla went to work for the Continental Edison Company in Paris. While on assignment in Strasbourg, France, in 1883, Tesla built his first inductor motor, one of the most common types of alternating-current motors. It was the first polyphase AC motor—meaning the coils in the motor were arranged in such a way that out-of-step currents energized the magnetic field, causing it to rotate at a predetermined speed. Such a discovery helped pave the way for a practical alternative to direct current. Alternating current was important because it could be more easily modified to suit a variety of situations.

Despite his highly technical way of thinking, Tesla was also a dreamer, prone to writing poetry when not working. True to his nature, in 1884, Tesla set out for America with a few poems and just four cents to his name. Upon arriving in America, Tesla worked for a brief time under the direction of Thomas Edison, but Edison was a strong proponent of direct-current technology and had little interest in the young engineer's work. Tesla quit, frustrated, after only one year.

In 1887, Tesla started the Tesla Electric Company in New York City. There, he experimented with a variety of technologies, including the shadowgraph, which presaged the invention of X-rays in 1895 by Wilhelm Röntgen. In 1888, Tesla sold his suite of AC inventions—the alternating-current motor and the transformers and devices that went along with them—to George Westinghouse. In 1893, Westinghouse used Tesla's system to light the World's Columbian Exposition in Chicago. Tesla's motors were used in 1893 to power the Niagara Falls Power Project. Tesla's humble AC motor harnessed the great power of the falls to deliver electricity 22 miles away in Buffalo, heralding a new era in modern electricity.

Tesla's other inventions include the 1891 Tesla coil, a high-frequency coil that is still used today in radio and television transmission equipment. He also experimented with remote control and radar, and developed fluorescent lighting. In fact, Tesla earned over one hundred patents in his lifetime, but he died nearly destitute, alone in his room at the famed New Yorker hotel. However, his legacy lives on, and in 1975 he was inducted into the National Inventors Hall of Fame.

Roller Coaster

Patent Name: Roller Coasting Structure
Patent Number: 310,966
Patent Date: January 20, 1885
Inventor: La Marcus A. Thompson, of South Chicago, Illinois

What It Does

This thrill ride, in its many variations, carries passengers along a track in small individual train-like cars, over steep drops and around precipitous curves, providing terror and delight in equal measure.

Background

Considerable advances in roller-coaster technology have focused on improving safety while maximizing thrill. Opened in 2005, Kingda Ka at Six Flags Great Adventure in Jackson, New Jersey, has the largest drop—418 feet—of any coaster to date.

In August 2007, Richard Rodriguez broke the record for the longest marathon on a roller coaster, 405 hours and 40 minutes.

In June 1884, the first contemporary commercial roller-coaster ride, the Gravity Pleasure Switchback Railway, opened to the public at Coney Island, New York. The ride cost a nickel a person and, at its fastest, moved at a whopping six miles per hour. It consisted of a platform for patrons, which included a ticket office, and flat steel tracks nailed into wooden planks, the whole of which was suspended at varying heights by trestles. Like many innovations, the "roller coasting structure" represented a combination of evolving concepts of the time. Elsewhere, others were devising similar pleasure rides, including boat-slide chutes and inclined railways, as the novelty of large-scale amusement parks was quickly becoming an industry.

It has been reported that L. A. Thompson was a preacher whose initial intention in building the ride was to divert attention from Coney Island's beer gardens. Preacher he may have been, but entrepreneur he quickly became. After enjoying the immediate success of his invention, Thompson went on to build dozens more roller coasters of varying designs. By the early 1900s, he had employed as a chief engineer a man named John Miller, who began designing his own rides. Miller held over one hundred patents on roller-coaster innovations and is today considered the father of the modern roller coaster. During the 1920s, roller coasters were all the rage, but following the Great Depression, the industry suffered enormously. In the four decades that followed, many more roller-coaster rides were dismantled than built. It is only recently, with the rise of such super-parks as the Great Adventure chain, that the roller race has heated up again, as engineers and designers strive to create the biggest, steepest, fastest, and most tortuous ride yet.

Coney Island's Flip Flap, built in 1895, was the first roller coaster with a loop.

How It Works

1. A parallel double track structure (B, B[1]) is supported by a trestle system (C), and includes both ends situated at an equal height above the rest of the track.
2. From a starting point (D) of the outgoing track, a car gains enough momentum as it descends to reach the other side.
3. Here, the car is transferred to the return track (B[1]) via a switch-track mechanism (E), and similarly moves back to its starting point.

In the Inventor's Words

"This construction and arrangement affords a very enjoyable means for amusement and pleasure, the sensation being similar to that of coasting on the snow, with the difference that the conveyance runs on wheels and returns the passenger to the starting point without the necessity of having to walk up hill for a second ride."

A roller skating rink in Washington, DC in 1880

Roller Skates

Patent Name: Roller Skate
Patent Number: 906,281
Patent Date: December 8, 1908
Inventor: James Leonard Plimpton, of Boston, Massachusetts

What It Does

First fashioned in 1863, the four-wheel roller skate gave skaters more maneuverability than its in-line predecessors, which were modeled on ice skates. Though it had begun as a recreation for the aristocracy, roller-skating became a popular and widespread amusement.

Background

Originally, roller skates were a lot more like today's in-line skates, roughly replicating the blade of an ice skate. According to Mary Bellis, who has been writing about inventors for years, a German ballet, *Der Maler oder die Wintervergnügen (The Artist, or Winter Pleasures)*, used roller skates to mimic ice-skating in 1818. The following year in France, a patent was issued for a skate that had in-line wheels.

James Plimpton changed the "in-line" design with his 1863 invention of a rocking skate, mounting two pairs of wheels side by side to allow skaters more maneuverability. Skaters could lean one way or another to steer as opposed to lifting their feet numerous times in increments. The innovation was favorably received, and skating continued to grow in popularity. An enterprising businessman, Plimpton would soon establish a skating club in New York City. This invention represents an improvement on the first patent awarded to the person who is now known as the father of modern roller skating.

How It Works

1. In figure 1, a shoe is secured to the skate by straps.
2. The trucks of the skate are designed to allow for leveraging to one side or another under pressure, allowing the direction of the skater to curve slightly in a preferred direction.

3. Braking is performed by applying pressure down on the heel where a screw bolt (48) engages the brake mechanism (47).

WITNESSES

INVENTOR
James Leonard Plimpton

BY

ATTORNEYS

In the Inventor's Words   "This invention relates to and is dependent upon the construction, arrangement, and mode of operation of that class of guidable curved-running roller skates, as originated and first patented by me in the United States, January 6, 1863, and more particularly as improved and patented by me in England, August 25, 1865, and in the [U]nited States, June 26, 1866. In this class of guidable curved-running roller skates the rollers are applied to the stock or foot-stand of the skate so that the said rollers may be cramped or turned, so as to cause the skate to run in curved lines either to the right or left by the turning, canting, or tilting laterally of the stock or foot-stand."

Slinky

Patent Name: Toy and Process of Use
Patent Number: 2,415,012
Patent Date: January 28, 1947
Inventor: Richard T. James

What walks down stairs, alone or in pairs, and makes a slinkity sound?
A spring, a spring, a marvelous thing. Everyone knows it's Slinky…
—TV commercial jingle

What It Does

The Slinky doesn't really do much—but it does it with endless style. The twangy little spring toy can "walk" downstairs, but mainly it just begs you to play with it. Kazhinnnnng.

Background

In 2001, the Slinky was inducted into the Toy Industry Hall of Fame. In the same year, it was named the official state toy of Pennsylvania.

The Slinky has been used as more than just a toy: Science teachers have used it as an educational aid, and in fighting the Vietnam War, soldiers tossed Slinkys into tree branches as makeshift radio antennae.

As a naval engineer in Philadelphia, Pennsylvania, Richard James experimented with tension springs in attempting to devise an antivibration system for sensitive equipment in ships. While he worked one afternoon in 1943, a coiled set of springs fell to the floor. The serendipitous accident revealed its unique potential outside of a ship—as a toy. To the engineer, it was the intriguing expression of potential energy transferable into kinetic energy. To the rest of the world, it was a coiled spring that, once set into motion, became instantly fascinating. It oscillated; it turned on its ends; it walked itself down stairs.

He and his wife, Betty, worked on fine-tuning the toy, and during the 1945 Christmas season they demonstrated the Slinky for the first time at a department store. In ninety minutes, the entire stock sold out, four hundred Slinkys for a buck apiece. Encouraged, the couple established a company to mass-produce the toys. James developed a machine that could turn 80 feet of steel wire—the length of a standard Slinky—into the desired coil in about ten seconds. James Industries was founded in 1956.

By the early 1960s, Slinky sales had slowed down. Richard James went off to join a religious group in Bolivia, leaving his wife with their six children and a lot of debt. Like a coiled spring, Betty James bounced back into motion. She took over at James Industries and saved Slinky from the shadows of obscurity, and the Slinky began to slink its way back into homes everywhere. Today more than 250 billion Slinkys have been sold worldwide.

How It Works

When set in motion on a stepped platform such as a stairway, a helical spring transfers energy along its length in a longitudinal wave. Coil by coil, and step by step, the whole spring descends end over end, as if it is somersaulting down one step at a time.

Jan. 28, 1947.   R. T. JAMES   2,415,012
TOY AND PROCESS OF USE
Filed Aug. 21, 1946   3 Sheets—Sheet 3

Fig.5

Fig.7

INVENTOR.
Richard T. James
BY
Attorneys.

In the Inventor's Words

"A purpose of my invention is to provide a helical spring toy which will transfer its turns from one end to the other in an entertaining manner when it is bent into general semi-circular form and the ends are moved up and down.

"A further purpose is to provide a helical spring toy which will walk on an amusement platform such as an inclined plane or set of steps from a starting point to successive lower landing points without application of external force beyond the starting force and the action of gravity."

Snowboard

Patent Name: Snowboard
Patent Number: 4,165,091
Patent Date: August 21, 1979
Inventor: Daniel E. Chadwick, of Rutherford, New Jersey

What It Does

The snowboard combines the principles of the surfboard, skateboard, and snow skis to offer a recreational glide over snow-laden surfaces.

Background

In 2015, Edmond Plawczyk set a world record for snowboarding speed—126 mph, beating the 1999 record by fractions of a second.

The popularity of snowboarding may seem like a recent phenomenon, but people have been enjoying the sport, in one form or another, for a long time. Sled-surfing was destined to evolve, and the flat-bottomed snowboard is its current manifestation. The skills and coordination required for snowboarding are often compared to those of skateboarding; and many street skaters answer the wintertime call of shredding the slopes.

How It Works

This patent represents an early version of the snowboard, which very closely resembles a modified skateboard and borrows many of the same underlying principles. The board consists of an elongated main body having laterally spaced skis or runners mounted to the underside of the body at the front and rear. The runners, like wheels on a skateboard, are coupled together in order to maintain their parallel relation and the stability of the board. Each runner may have a centrally disposed stabilizer fin on its bottom surface.

1. A board is constructed from rigid plastic or wood.
2. Axially spaced pairs of runners with upward-curving portions are situated on the bottom, near both ends.
3. The runners feature small central fins, which help stabilize the board while in use.

4. The runners are symmetrically located on opposite sides of the board's longitudinal axis and coupled together in parallel relation by transverse bars.

5. A middle transverse bar is vertically offset above the front and rear bars, and includes a flexible bracket, the upper end of which is secured to the bottom of the board.

U.S. Patent   Aug. 21, 1979   4,165,091

In the Inventor's Words "According to a preferred mode of operation the board is used by placing it at the top of a hill which preferably is packed hard with snow. The operator stands on the board and pushes off, whereupon both feet are placed on the board, being located similar to their location on a 'skateboard.' As the board slides over the snow proceeding downhill, the operator may lean to either side in order to manipulate the board in a preferred direction. The resilient connection will aid in manipulation of the board by the operator."

Auguste Bartholdi in his workshop with the inner framework of the statue

Statue of Liberty

Patent Name: Design for a Statue
Patent Number: 11,023
Patent Date: February 18, 1879
Inventor: Auguste Bartholdi, of Paris, France

What It Does

Offered as a gift by France to the United States, the monumental Statue of Liberty, which sits in New York Harbor, was designed to embody Liberty enlightening the world.

Background

Joseph Pulitzer helped obtain contributions to complete the statue's base in April 1886. Benefit funds for Lady Liberty were also raised by theatrical events, art exhibitions, auctions, and prizes.

She is a tourist attraction; she is a piece of art; she is a beacon of welcome to America's immigrants enchanted by the possibilities of a better life in a new land. One of the single most photographed objects in the world, Lady Liberty is a universal symbol of freedom. A gift from France to commemorate the centennial of the American Declaration of Independence, the statue was conceived at a dinner party in Paris in 1865. Six years later, in order to check out sites and get some ideas for the monument, the French sculptor boarded a ship bound for America. Before it docked, Bartholdi had selected Bedloe's Island as the site and had even drawn up some sketches.

In 1876, the statue's torch-bearing arm was demonstrated in Philadelphia in honor of the centennial of the Declaration of Independence; the rest of the statue had not yet been finished. The sculptor conceived of the whole of his work as consisting of an inner supporting framework of steel covered by plates of thin copper. Alexandre Gustav Eiffel, who later built another famously photogenic monument, oversaw construction of the inner framework. Completed in France in July 1884, the statue arrived the following year by boat, like so many European immigrants of the day. The vessel carried Lady Liberty in more than three hundred pieces packed in more than two hundred crates. The statue was inaugurated by President Grover Cleveland on October 28, 1886. In 1903, these verses of Emma Lazarus were inscribed at the statue's base:

Give me your tired, your poor,
your huddled masses yearning to breathe free,
the wretched refuse of your teeming shore.
Send these, the homeless, tempest-tost to me.
I lift my lamp beside the golden door!

How It Works An inner framework of steel is sheathed in multiple molded plates of thin copper.

STATUE STATS

- Height from base to torch:
  151' 1" (46.05m)
- Length of right arm:
  42' (12.8m)
- Index finger:
  8' (2.44m)

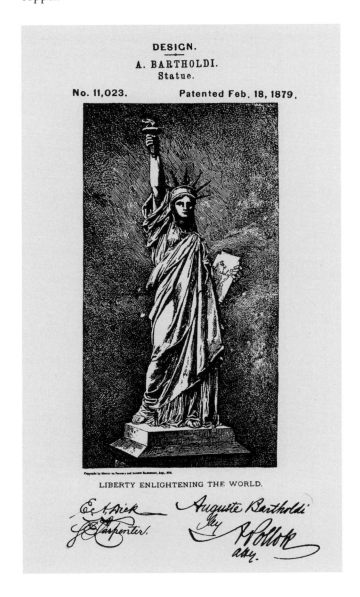

DESIGN.

A. BARTHOLDI.
Statue.

No. 11,023.   Patented Feb. 18, 1879.

LIBERTY ENLIGHTENING THE WORLD.

In the Inventor's Words

"The statue is that of a female figure standing erect upon a pedestal or block, the body being thrown slightly over to the left, so as to gravitate upon the left leg, the whole figure being thus in equilibrium, and symmetrically arranged with respect to a perpendicular line or axis passing through the head and left foot. The right leg, with its lower limb thrown back, is bent, resting upon the bent toe, thus giving grace to the general attitude of the figure. The body is clothed in the classical drapery, being a stola, or mantle gathered in upon the left shoulder and thrown over the skirt or tunic or under-garment, which drops in voluminous folds upon the feet. The right arm is thrown up and stretched out, with a flamboyant torch grasped in the hand. The flame of the torch is thus held high up above the figure. The arm is nude; the drapery of the sleeve is dropping down upon the shoulder in voluminous folds. In the left arm, which is falling against the body, is held a tablet, upon which is inscribed '4th July, 1776.'"

Television

Patent Name: Cathode Ray Tube
Patent Number: 2,139,296
Patent Date: December 6, 1938
Inventor: Vladimir K. Zworykin, of Philadelphia, Pennsylvania, assignor to Radio Corporation of America

What It Does

The television is the bearer of news, the provider of entertainment, and a medium for commercial advertising.

Background

The story of the invention of television would itself make for a compelling TV drama. To this day, there remains debate as to who should

Within seconds after it was introduced, the TV set moved from the
living room into the kitchen...and
the bedroom, and the kid's room,
and finally even the car...

receive credit. But like so many complex innovations, the TV was
an awesome creation just waiting to happen, a keystone balanced
atop previous ingenious ideas. One of the innovations key to its
foundation—the cathode ray tube—was invented in 1897 by a German
scientist named Karl Braun. A specialized vacuum tube that can transfer
refracted electron beams into images, the cathode ray tube is to this day
used in TVs, video cameras, and many other image-related technologies.

About a decade after Braun created his cathode ray tube, a Russian
physicist named Boris Rosing began tinkering with the idea, enlisting
the help of his student Vladimir Zworykin. While Rosing was first to
use Braun's tube in an attempt at producing television images, he could
only manage to transmit crude shapes. Rosing's experiments were cut
short when he disappeared sometime during the 1917 Bolshevik Revolution. Eventually, his student would enjoy another chance at realizing
the dream.

Zworykin left his home country to study the relatively new X-ray
technology in Paris, eventually traveling to the United States in 1919,
where he got a job at a Westinghouse lab in Pittsburgh, Pennsylvania.
Here, and later with significant funding from RCA, he continued to
experiment with TV technology, making significant improvements to
the cathode ray tube for the purpose of generating electronic images.

RCA would later fund another inventor who made a legitimate case
that he had already invented and patented the basic components of
electronic television. A farm boy from Utah, Philo Farnsworth was the
first to transmit an image made by sixty horizontal lines on a TV screen.
In 1927, he filed for his first television patent, number 1,773,980. RCA
sent its Russian engineer to meet the Utah farm boy to obtain information on perfecting its television. Farnsworth's company hosted the
Russian, thinking it might be able to obtain a valuable contract from
the much-larger RCA.

After arduous legal wranglings and patent standoffs, RCA conceded
to pay $1 million for the rights to Farnsworth's patents. Though this
fact is probably not something RCA would advertise in its corporate
history, it still does not answer with total certainty the question of who
was the true television inventor, if there is only one, and debate on the
subject lingers. Both Zworykin and Farnsworth had workable television
systems by the year 1930, and both are credited as "the inventor."

What is certain: Television has made the world smaller and has profoundly altered social landscapes everywhere.

How It Works

1. Within the cathode ray tube, electrons are directed at a photosensitive
   target, and are refracted toward its flat surface, which comprises the
   television screen.

2. When electrons strike the phosphorescent coating along the flat surface
   of the tube, light is emitted.

3. Meanwhile, around the outside of the tube, electromagnetic coils deflect the electron beam so that it scans across the screen, usually in a horizontal pattern.

4. When the scanning is maintained at a rapid pace, the light on the screen forms images that appear steady and continuous to the human eye.

Dec. 6, 1938.   V. K. ZWORYKIN   2,139,296
CATHODE RAY TUBE
Filed Feb. 23, 1934

Fig.1.

Fig.2.   Fig.3.

INVENTOR:
Vladimir K. Zworykin
BY R. R. Goldsborough
ATTORNEY.

In the Inventor's Words   "A cathode ray transmitting tube as disclosed in my copending application is constituted by an evacuated container having at least one transparent wall, a planar photo-sensitive cathode of the mosaic type, and an electron gun for directing a cathode ray toward and against the cathode.

"When such a device is utilized in a television transmitter or in an ultra-microscope of the type disclosed in the aforesaid application, it is necessary to interpose an optical system, constituted by one or more lenses, between the object and the photo-sensitive electrode for the purpose of forming upon the said electrode an optical image of the object."

Virtual Reality

Patent Name: Method and Apparatus for Creating a Wireframe and Polygon Virtual World
Patent Number: US 5559995 A
Patent Date: September 24, 1996
Inventors: Dan D. Browning, Ethan D. Joffe, and Jaron Z. Lanier, of San Francisco, California

What It Does

This head-mounted display device gives users the illusion of being in a virtual world and allows them to navigate through the virtual landscape with either sensors on the headset or a handheld controller.

Background

Timothy Leary, psychologist and LSD advocate, got in touch with VPL Research in an attempt to find a legal way to study consciousness that would imitate taking psychedelic drugs.

In 1984, Jaron Lanier, the computer scientist, artist, and musician who is generally credited with coining the term "virtual reality," or VR, founded VPL Research. Lanier originally hoped to develop a VPL, or "visual programming language," that would allow more people to learn how to program computer systems. VPL Research went on to become one of the very first companies to develop and sell virtual-reality devices.

One of VPL Research's first devices was the DataGlove. Invented by Lanier's partner at VPL, Thomas Zimmerman, the glove was originally intended as an input and control method for computers. After bringing on Mitch Altman to help program the microcontrollers of the glove, it became apparent that the DataGlove would have applications in a virtual-reality system.

The next major piece of hardware from VPL Research was the EyePhone, a head-mounted display (HMD) similar to the virtual-reality headsets seen today. The headset was designed to immerse users in a virtual world, using LCD screens with variations in the images to create a sense of depth. VPL would go on to develop bodysuits, audio systems, and 3D-rendering engines to refine its virtual-reality system.

Lanier and VPL Research were issued a number of patents related to VR technology. One of the first, referenced here, describes an immersive operating system where users actively program and shape the world around them, all in VR. Instead of putting on a headset and seeing, say, a mountain landscape or a video game world, users needed to operate input controls to program the world around them before any colored or textured objects would appear before them. True to the early goal of VPL Research, Lanier wanted to bring computer programming to a wider population and "make virtual world creation as simple as possible."

Virtual-reality technology became widespread in spring of 2016, when the Oculus Rift headset was released on March 28 and the HTC Vive was released just eight days later. These two VR headsets, along with various smartphone-based setups like Samsung Gear VR and Google Cardboard, popularized virtual reality within the consumer tech market. Today, video games and other entertainment are developed specially for VR, and national sports leagues such as the NFL and the NBA offer VR viewing for games. Augmented reality, or AR, the technology used in products like Microsoft HoloLens, shows you the real world with an overlay of computer-generated information. Both VR and AR are being integrated into many industries, from marketing to construction to NASA science missions.

How It Works

1. A head-mounted display (HMD), generally with one or more LCD screens, is connected with an HDMI cable to a computer that runs software to generate a three-dimensional virtual world. If the VR device is a smartphone, the phone simply runs the software.

2. The headset tracks where the user is looking with a series of sensors, and displays the appropriate portion of the virtual world on the screen.

3. To immerse the user and mimic reality, the display shows about a 100-degree field of view, slightly less than the human eye is capable of.

4. Additional lenses in the headset create a stereoscopic 3D image that mimics the slight variations in each eye's field of view.

5. The user can navigate through the virtual world and interact with objects in it simply by moving, as tracked by sensors in the headset, or by using handheld controllers, which incorporate sensors with control systems such as joysticks and buttons.

In the Inventors' Words

"When children are growing up, they face a profound conflict between the internal world of their dreams and imagination, in which everything's possible and fluid, and the practical world in which they have parents, food, and friends, in which they're not alone, and in which they can survive. So as kids grow up, they have to gradually de-emphasize this world of imagination and celebration and emphasize the practical world, unless they're willing to be alone in their insanity and completely dependent on others for survival. Of course it's possible to integrate the two, but it's so hard, like walking a tightrope. I think the reason that kids instinctively love computers, and especially love virtual reality, is that it really does present a new solution, a way to make imaginary worlds that we can be together in, just like the real world." —Jaron Lanier

Seven

Resources

Bibliography

Websites

Resources

THE THREE TYPES OF PATENTS

Three types of patents are awarded by the United States Patent and Trade Office. The most commonly issued patent, by far, is the utility patent. The other two types are plant patents and design patents.

1. Utility Patents

Utility patents are issued to any invention that can be considered a useful process, machine, method, or composition of matter. Utility patents also cover the development of new uses for any previously existing inventions in any of these categories, as well as improvements upon inventions that are already around. The good thing, as far as inventors are concerned, is that patents don't have to be more useful, original, or appealing than anything else on the market; they simply have to have a baseline of usefulness about them. The key to being granted a utility patent lies in the functionality of the invention. As long as something can be considered "new and nonobvious" (in the words of the patent office), it will most likely be awarded a patent.

The extreme majority of patents represented in this book, as in life, are utility patents. The field of biotechnology has broadened the definition of the utility patent, which are now granted to cover forms of life developed in a laboratory through genetic engineering. New breeds of mice, never-before-seen strains of corn, and many types of microorganisms are examples of the utility patent. Utility patents expire twenty years after being filed.

2. Plant Patents

The Plant Patent Act of 1930 gave intellectual property protection to inventors of new kinds of plants. Plant patents are granted to anyone who discovers or invents and asexually reproduces a distinct and new variety of plant. This includes mutants, hybrids, or newly found seedlings. An asexually reproduced plant has been propagated not from a seed but from cutting, grafting, inarching, or another similar process. Plant patents do not apply to tuber-propagated plants, such as the Irish potato or Jerusalem artichoke. When submitting a patent application, the inventor of the new plant must include a detailed drawing of the specimen; if the new plant is distinguished by a new color, the color must be shown in the drawing. Plant patents are often issued to individuals who have created new plants for a variety of reasons: to create trees that bear more fruit, or a new kind of fruit; to create plants (or crops) that are more resilient to disease; to cultivate plants with medicinal properties; or simply to bring into existence a new kind of aesthetically pleasing flower. Like utility patents, plant patents expire after twenty years.

3. Design Patents

Design patents are granted to anyone who has come up with a new and "nonobvious" ornamental design for a manufactured item. One such example in this book is the Fender Stratocaster. The design patent prevents anyone else from making something that resembles the product. Approximately one design patent is issued for every ten utility patents out there, making them relatively rare in the patent world. The interesting thing about design patents is that they only cover the cosmetic appearance of a thing—the patent has nothing whatsoever to do with the item's function. As a result, when someone files a design patent application, the entire "invention" can only be illustrated through a drawing or other rendering; it is not considered valid if it is described in words. Design patents are sought out when the appearance of a thing will substantially influence that item's commercial success. So, for example, computer icons are often patented; however, the parts inside of a computer, which aren't normally seen, cannot be issued design patents, because it isn't the shape—or design—of the parts that makes them functionally useful. Design patents expire after fourteen years.

PATENTS VERSUS TRADEMARKS

Patents and trademarks form the basis of the work of the U.S. Patent and Trademark Office. Both protect holders of certain types of intellectual property. However, the exact function of the each is distinct.

A patent is a document that protects an invention, usually a product, plant, or design (see "The Three Types of Patents"). When the inventor is issued the patent, he or she receives certain protections that allow others to use, analyze, and/or modify the invention in the name of the advancement of science, while still granting the inventor any proceeds, profits, or other benefits that may arise from the invention.

On the other hand, there are trademarks: Bubble Wrap; Post-it notes; Etch A Sketch; Velcro. Each of these creations is revolutionary in its own way; each has its story to tell; and each is trademarked, which is why you see the little circle with the "R" after each name. That symbol means "registered trademark." The success of these creations relies on brand-name recognition—and on successfully establishing the fact that the product is identifiable, distinct, and superior. While patents protect the duplication of a product by another manufacturer for a limited time, trademarks rely on consumer devotion to the original product, through marketing its name, next to which other similar products then become measured. Unlike patents, a trademark is valid for as long as the product is in use.

A trademark is simply a word, phrase, symbol, or design that appears on the product being sold and which distinguishes itself from other words, phrases, symbols, or designs in the marketplace. A service mark, yet another type of trademark, is used by companies that provide a service instead of a product.

A service mark is also represented by a word, phrase, or symbol. Another word for trademark is "brand name." Companies work hard to make their trademarks distinctive. A strong trademark may become so embedded in the public's consciousness, in fact, that it virtually replaces the real name of the product. The above listed products serve as outstanding examples. One typically doesn't ask for "sealed-air cushioning material" when "Bubble Wrap" is so ingrained in the lexicon. "Kleenex" and "Xerox" offer other similar examples. These aren't vocabulary words at all, but trademarks or brand names. Same goes for Coke, the nickname for Coca-Cola. But while the Coca-Cola name is trademarked, the formula used to create its flavor has been kept as a trade secret.

In other words, a trademark really protects a marketing idea—the image of a product that a company is trying to sell. When a company issues a trademark, it wants to make sure that the consumer doesn't confuse its product with some other company's product. So the trademark protects the company's image or reputation and, in doing so, helps it build brand loyalty and stay in business. Trademarks have become increasingly valuable as the global marketplace has expanded. In a market flooded by goods from all over the world, companies are challenged to work harder to establish and protect their trademarks.

HOW TO OBTAIN A PATENT

The first thing you want to do before applying for a patent is to make sure that your idea hasn't already been patented. You can check on this by conducting a literature review at your local university library, or by searching on the U.S. Patent and Trademark Office website (http://www.uspto.gov/). While you're putting the final touches on your invention, be secretive! It's a good idea to keep records and documents of all the work, purchases, and other data that went into the making and refining of your idea, just in case someone else comes up with a similar idea before you apply for your patent. That way, if a similar invention is unveiled around the same time, you can prove that you indeed had the idea first.

Once you're pretty sure that your invention is "new and nonobvious," you can begin the patent application process. There are three main components of any patent application: (1) a written document that specifies the details of the invention and includes an oath of declaration; (2) an accompanying drawing of the invention, if that's applicable; and (3) the filing fee. The U.S. Patent and Trademark Office calls the patent application a "complex legal document" that is best left to patent attorneys or agents. However, you can apply for one on your own if you're willing to fulfill the lengthy and exacting requirements.

Once you've attended to the details, the bulk of the application can be found in the specification section. This section requires a full, concise, clear, and exact description of the invention that is explicit enough so that another person could construct or replicate it. Then comes all-important Claims section. Here, you must make as strong a case as possible for the relevance and usefulness of your invention, citing the aspects that make it distinct. The wording here is extremely important and strongly influences whether your patent will be granted.

Next is the Drawing section, which should include various views of the invention, if applicable. Finally, you must be sure to pay the applicable filing fees. If your patent is granted, congratulations! But be prepared to pay even more money, including an issuance, design, and plant fee. You will also have to pay various maintenance fees at 3½ years, 7½ years, and 11½ years after issuance of the patent. The maintenance fees grow incrementally as the years go by, but if your patent has proved useful to society, you'll surely be more than rich enough to afford them!

USEFUL RESOURCES

You can search for U.S. Patents in the Search Room of the Patent and Trademark Office in Crystal Plaza, 2021 Jefferson Davis Highway, Arlington, Virginia. Or, you can search for U.S. patents on the Internet at: http://www.uspto.gov/patft/index.html. You can also call the USPTO's General Information Services Division at 800-PTO-9199 or 703-308-4357.

In lieu of traveling to Virginia to manually search for patents, you can make a trip to your nearest Patent and Trademark Depository Library (PTDL). Every state as well as the District of Columbia and Puerto Rico have PTDLs, where you can search patent and trademark collections. To find the nearest

PTDL, visit http://www.uspto.gov/web/offices/ac/ido/ptdl/ptdlib_1.html.

Online instructions for filing patent applications can be found on the United State Patent and Trademark Office, at these addresses:

A Guide to Filing a Utility Patent Application
http://www.uspto.gov/web/offices/pac/utility/utility.htm#intro

A Guide to Filing a Design Patent Application
http://www.uspto.gov/web/offices/pac/design/index.html

General Information About 35 U.S.C. 161 Plant Patents
http://www.uspto.gov/web/offices/pac/plant/index.html

You can also file for some applications online, such as utility patents and provisional applications, using the USPTO's electronic filing system. You cannot file online for design or new plant applications.

Electronic Filing System
www.uspto.gov/patents-application-process/file-online

There are lots of fun sites out there dedicated to patent news, laws, history, and information. Here are some good ones:

National Inventors Hall of Fame
A fun, nicely designed site whose aim lies in "fostering the inventive spirit in all of us."

http://www.invent.org

Patent-FAQ
This list of frequently asked questions assembled by registered U.S. patent agent David Kiewit is a treasure trove of information regarding patent applications and intellectual property in general.

http://www.patent-faq.com/

Patent Law
For comprehensive information about patent laws, visit the American Patent and Trademark Law Center at:

http://www.patentpending.com/

Bibliography

Amirani, Amir. "Sir Alec Jeffreys on DNA Profiling and Minisatellites," *Science Watch*, 1996.

Bacon, Tony. *The Ultimate Guitar Book*. New York: Alfred A. Knopf, 1997.

Bardey, Catherine. *Lingerie: A History and Celebration of Silks, Satins, Laces, Linens and Other Bare Essentials*. New York: Black Dog and Leventhal, 2001.

Bowler, R. M., PhD, MPH, and J. E. Cone, MD, MPH. *Occupational Medicine Secrets*. Philadelphia, PA: Hanley & Belfies, 1999.

Bullock, A., and R. B. Woodings, eds. *20th Century Culture*. New York: Harper & Row, 1983.

"Chili Peppers and Endorphins," *The Veiled Chameleon*, April 21, 2003, http://www.veiled-chameleon.com/archives/000042.html (accessed November 11, 2003).

Folkhard, Claire, ed. *2003 Guinness Book of World Records*. New York: Bantam Books, 2003.

Folkhard, Claire, ed. *2004 Guinness Book of World Records*. New York: Time, Inc., 2003.

"Genetic Fingerprinting," *The Science Show*, Radio National. Transcript of Broadcast, September 21, 2002.

Hancock, Michael. "Burroughs Adding Machine Company: Glimpses into the Past," http://www.dotpoint.com/xnumber/hancock7.htm (accessed September 2003).

Harrison, I., and S. Fossett. *The Book of Firsts: The Fascinating Stories Behind the World's Greatest Achievements, Discoveries, and Breakthroughs*. Pleasantville, NY: Reader's Digest, 2003.

I'll Buy That! 50 Small Wonders and Big Deals That Revolutionized the Lives of Consumers. Yonkers, NY: Consumers Union, 1986.

Kemmiya, Misa. "TED Case Studies, Shrimp and Turtle, Case Number 436," http://www.american.edu/projects/mandala/ted/shrimp2.htm (accessed September 2003).

Kwolek, Stephanie. Telephone interview by author. August 24, 2003.

Lillard, Margaret. "Re-Enactment of Wright Bros. Flight Fails," Associated Press, December 17, 2003.

McLaren, Carrie. "Porn Flakes: Kellogg, Graham and the Crusade for Moral Fiber," ibiblio: the public's library and digital archive, http://www.ibiblio.org/stayfree/10/graham.htm (accessed October 2003).

Mooney, Julie. *Ripley's Believe It or Not! Encyclopedia of the Bizarre, Amazing, Strange, Inexplicable, Weird and All True!* New York: Black Dog and Leventhal, 2002.

Papazian, Charlie. *The Complete Joy of Home Brewing*, 3d ed. New York: Harper Collins, 2003.

Rose, S., and N. Schlager. *How Things Are Made*. New York: Black Dog and Leventhal, 2003.

Schiavone, Louise. "Australian Inventor's Gun Fires 1 Million Rounds a Minute," CNN, June 28, 1997, http://www.cnn.com/TECH/9706/28/super.gun/ (accessed October 2003).

Turkington. C. A., and S. J. Propst, MD. *The Unofficial Guide to Women's Health*. Boston, MA: IDG Books Worldwide, 2000.

Van Dorenstern, Phillip, ed. *The Portable Poe*. New York: Penguin Books, 1981

Websites

Access Cash
http://www.access-cash.com

American Plastics Council
https://plastics.americanchemistry.com/

Band-Aid Brand Adhesive Bandages
http://www.band-aid.com

British Airways
http://www.britishairways.com/concorde/index.html

Capt'n Clint's Place
http://capt.clint.home.mindspring.com/index.html

Clonaid Pioneers in Human Cloning
http://www.clonaid.com/news.php

DEKA Research and Development Corporation
http://www.dekaresearch.com

Devil's Rope Museum
http://www.barbwiremuseum.com/

Dupont
http://www1.dupont.com

Edward Lowe Foundation
http://edwardlowe.org

Enchanted Learning
http://www.zoomschool.com/Home.shtml

California Energy Commission, Energy Quest
http://www.energyquest.ca.gov

Carrier Corporation
http://www.carrier.com/

Drive-ins.com: The Definitive Resource for Drive-In Information
http://www.drive-ins.com

Etch-A-Sketch.com
http://www.etchasketch.com

Fender.com
http://www.fender.com

Goodyear
http://www.goodyear.com/

Forbes
http://www.forbes.com/

How Stuff Works
http://www.howstuffworks.com/

Goddard Space Flight Center, NASA
http://www.gsfc.nasa.gov/

The Great Idea Finder
http://www.ideafinder.com/

The History of Computing Project
http://www.thocp.net/index.htm

ibiblio: The Public's Library and Digital Archive
http://www.ibiblio.org

Inventors with Mary Bellis, at About, Inc.
https://www.thoughtco.com/inventions-4133303

John Deere
http://www.deere.com

Joseph Enterprises, Inc.
http://www.chia.com/

Kellogg's
http://www.kelloggs.com

Kentucky Educational Television
http://www.ket.org/

Lemelson-MIT Program
http://web.mit.edu/invent/

Levi Strauss & Co.
http://www.levistrauss.com/

The Library of Congress
http://www.loc.gov/

Lucidcafé
http://www.lucidcafe.com

Museum of Talking Boards
http://www.museumoftalkingboards.com/

National Fire Protection Association
http://www.nfpa.org/catalog/home/index.asp

The National Museum of American History,
Smithsonian Institution, Fast Attacks and Boomers:
Submarines in the Cold War
http://americanhistory.si.edu/subs/index.html

National Oceanic and Atmospheric Administration
http://www.noaa.gov/

National Park Service, Statue of Liberty
http://www.nps.gov/stli/

The Nobel Prize
http://www.nobelprize.org

Rothschild Petersen Patent Model Museum
http://www.patentmodel.org/default.aspx

PEZ Candy
http://www.pez.com/

Prozac.com
http://www.prozac.com/

Recovery Pharmaceuticals, Inc.
https://dailymed.nlm.nih.gov/dailymed/index.cfm

The Simon Lake Submarine Website 2003
http://www.simonlake.com

Clarivate Analytics
http://ip.thomsonreuters.com/maketheswitch

3M Worldwide
http://www.3m.com/

The Toxic Shock Syndrome Information Service
http://www.toxicshock.com/

UK Intellectual Property Office
http://www.patent.gov.uk/

United States Patent and Trademark Office
http://www.uspto.gov/

University of New Mexico School of Medicine
Department of Anesthesiology and Critical Care
Medicine
http://anesthesiology.unm.edu/

U.S. Food and Drug Administration
http://www.fda.gov

Viagra.com
http://www.viagra.com/

Vinyl Record Day
http://www.vinylrecordday.com/

Wikipedia, the Free Encyclopedia
http://en.wikipedia.org/

Wilkes-Barre Township Police Department
http://twp.wilkesbarre.pa.us/

Zenith Electronics Corporation
http://www.zenith.com/

Zippo Manufacturing Co.
http://www.zippo.com/home.html

Index

HTC Vive, 257
Huang, Wen C., 19, 20
Hull, Charles W., 50
Humayun, M., 111
Hyperloop Transportation
Technologies, 157

I

IBM (International Business
Machines Corporation), 91,
119
iBot, 113
Independence Technology, 114
Intel 8080, 92
Internal combustion engine,
23–26
International Business Machines
Corporation (IBM), 91, 119
iPad, 28, 93
iPhone, 27–29, 93, 209
iPod, 28, 29
ITU-R (International
Telecommunication Union
Radiocommunication Sector),
54
Ive, Jonathan P., 27

J

Jackson, Andrew, 146
Jacob, Mary P., 111, 112
Jacquard, Joseph-Marie, 90
James, Betty, 247
James, Richard T., 247, 248
James Industries, 247
Jang, Bor Z., 19, 20
Jeans, Levi Strauss, 82–83
Jefferson, Thomas, xi
Jeffreys, Alec J., 14
Jennings, Thomas L., 216
Jobs, Steve, 27, 28, 92, 93
Joffe, Ethan D., 256

John Deere plow, 193–194
Johnson, W. H., 219
Johnson & Johnson, 100, 101,
114
Jones, Kelly E., 8
Juan, E. de, 111
Judson, Whitcomb L., 95
Jukebox, 232–233
Julian, Percy, 216

K

Kahlke, Henry W., 189
Kamen, Dean L., 113, 114, 116
Keck, Donald B., 83
Keeler, Leonard, 202
Kellogg, John, 224, 225
Kellogg, Will, 224
Kennard, Charles W., 160
Kennard Novelty Company, 161
Kennedy, John F., 36
Kerr, Duncan Robert, 27
Kerwin, John M., 113
Kettering, Charles F., 26, 38
Kevlar, 117–118
Key, Brian R., 113
Kitty litter, 194–195
Kleenex, 261
Knight, Phil, 152
Kochevar, Bill, 9
Kodak, 226
Kolff, William J., 4
Kraut, K., 64
Kurzweil, Raymond C., 130, 131
Kwolek, Stephanie, 117

L

Laithwaite, Eric, 156
Lake, Simon, 43, 44
Lake Submarine Company, 44
Lallement, Pierre, 176

Lamp, lava, 154–155
Lanier, Jaron Z., 256, 257, 258
Larson, John, 202
Laser eye surgery, 118–120
Latimer, Lewis, 216
Lauterbur, Paul C., 121
Lava lamp, 154–155
Lava Lite, 154
Lawn mower, 196–198
Lazarus, Emma, 252
Leg, artificial, 107–108
Leibniz, Gottfried Wilhelm, 90
Leland, Henry M., 25
Lenin, V., 166
Lennon, J., 146
Lenoir, Étienne, 25
Levi Strauss blue jeans, 82–83
Levitra, 220
Lewin, W. S., 109
Lie detector, 202–203
Lightbulb, 29–32
Lighter, Zippo, 97–98
Lilly, Eli, 220
Lock, for door, 183–185
Loughead, Malcolm, 26
Lowe, Edward H., 194, 195
Lowe, Henry, 195

M

Macintosh, 92
Maglev, 156–158
Magnetic Resonance Imaging
(MRI), 120–122
Mahon, Geoffrey L., 102
Mansfield, Peter, 121
Marconi, Guglielmo, 240
Martin, Steve, 125
Masters, G., 219
Mathias, John Paul, 219
Mauchly, John William, 91

Photo Credits

p. 2 Everett Historical / Shutterstock.com; p.4 Dariush M / Shutterstock.com; p.6 Shutterstock.com; p. 8 dpa picture alliance archive / Alamy Stock Photo; p. 10 Photographic History Collection, Smithsonian's National Museum of American History; Lee Boltin / Contributor / Alamy Stock Photo; p.12 Anna Webber Stringer / Getty Images; p.14 Andrey Kuzmin / Shutterstock.com; p. 15 Three Lions / Stringer / Getty Images; p. 16 Library of Congress; p. 17 Shutterstock.com; p. 18 U.S. Naval Research Laboratory; p. 25 World History Archive / Alamy Stock Photo, Library of Congress; p. 26 Library of Congress; p. 27 blackzheep / Shutterstock.com, Featureflash Photo Agency / Shutterstock.com; p. 29 Library of Congress; p. 31 Library of Congress; p. 32 Everett Historical / Shutterstock.com; p. 33 Marka / Alamy; p. 35 Everett Historical / Shutterstock.com; p. 37 Rattiya Thongdumhyu / Shutterstock.com; p. 38 James Steidl / Shutterstock.com; p. 40 Everett Collection Inc. / Alamy; p. 42 shutterlk / Shutterstock.com, Library of Congress; p. 46 Everett Historical / Shutterstock.com; p. 48 Everett Historical / Shutterstock.com; p. 50 Alexander Tolstykh / Shutterstock.com; p. 53 GaudiLab / Shutterstock.com; p. 55 Anna Baburkina / Shutterstock.com; p. 57 Library of Congress; p. 58 Cultura Creative (RF) / Alamy Stock Photo; p. 59 Shutterstock.com; p. 60 Shutterstock.com; p. 62 dimbar76 / Shutterstock.com; p. 64 Science History Images / Alamy Stock Photo; p. 65 Pictorial Press LTD / Alamy Stock Photo; p. 67 Coprid / Shutterstock.com; p. 69 Paul Fearn / Alamy Stock Photo; p. 70 bilwissedition Ltd. & Co. KG / Alamy Stock Photo; p. 74 Eric E Johnson; p. 76 African Studio / Shutterstock.com; p. 78 kleem26 / Shutterstock.com; p. 80 Library of Congress; p. 82 AlenKadr / Shutterstock.com. p. 84 Elsar / Shutterstock.com; p. 86 Media Drum World / Alamy Stock Photo; p. 90 Library of Congress; p. 91 Science History Images / Alamy Stock Photo, RGB Ventures / SuperStock / Alamy Stock Photo; p. 92 National Museum of American History, Behring Center Archives Center, Mahyuddin / Shutterstock.com, drserg / Shutterstock.com; p. 93 Joris van den Heuvel / Shutterstock.com; p. 95 ang intaravichian / Shutterstock.com; p. 97 givaga / shutterstock; p. 100 20th Century Advertising / Alamy Stock Photo; p. 102 Vereshchagin Dmitry / Shutterstock.com; p. 105 Science History

Images / Alamy Stock Photo; p. 107 Library of Congress; p. 109 Anadolu Agency / Contributor / Getty Images; p. 113 dbimages / Alamy Stock Photo; p. 116 Paul Morigi / Stringer / Getty Images; p. 117 Shutterstock.com; p. 119 Shutterstock.com; p. 120 Shutterstock.com; p. 123 Boston Globe / Contributor / Getty Images; p. 125 Shutterstock.com; p. 127 Shutterstock.com; p. 129 Cartoon Resource / Shutterstock.com; p. 131 dpa picture alliance / Alamy Stock Photo; p. 132 Shutterstock.com; p. 135 Shutterstock.com; p. 137 Shutterstock.com; p. 139 Science History Images / Alamy Stock Photo; p. 142 AP Photo / ED Kolenovsky; p. 144 Shutterstock.com; p. 151 PA Images / Alamy Stock Photo; p. 152 AlenKadr / Shutterstock.com; p. 154 Shutterstock.com; p. 156 Shutterstock.com; p. 159 Shtuterstock.com; p. 161 Shutterstock.com; p. 163 Shutterstock.com; p. 165 Topical Press Agency / Stringer / Getty Images; p. 170 Shutterstock.com; p. 174 Shutterstock.com; p. 178 Library of Congress; p. 182 IllustratedHistory / Alamy Stock Photo; p. 184 Historical / Contributor / Getty; p. 187 Bettmann / Contributor / Getty Images; p. 189 History Museum at the Castle; p. 191 Shutterstock.com; p. 194 Shutterstock.com; p. 198 Shutterstock.com; Pictorial Parade / Getty Images; p. 200 Shutterstock.com; p. 202 Shutterstock.com; p. 204 Minnesota Historical Society; p. 206 Ira Gay Sealy / Contributor / Getty Images; p. 208 Shutterstock.com; p. 210 Shutterstock.com; p. 212 Library of Congress; p. 214 MCNY / Gottscho-Schleisner / Contributor / Getty Images; p. 216 Science History Images / Alamy Stock Photo; p. 219 PureRadiancePhoto / Shutterstock.com; p. 222 Art Babych / Shuttertock.com; p. 224 Library of Congress; p. 225 Library of Congress; p. 226 J.D.S / Shutterstock.com; ullstein bild Dtl. / Contributor / Getty Images; p. 228 Chris Willson / Alamy Stock Photo; p. 230 dean bertoncelj / Shutterstock.com, Val Wilmer / Contributor / Getty Images; p. 232 Four Oaks / Shutterstock.com, Fotosearch / Stringer / Getty Images; p. 236 Library of Congress; p. 238 Dmitry Kalinovsky / Shutterstock.com; p. 240 Lebrecht Music and Arts Photo Library / Alamy Stock Photo; p. 242 GL Archive / Alamy Stock Photo; p. 243 shutterstock.com; p. 244 Library of Congress; p. 245 Library of Congress; p. 247 Bettmann / Contributor / Getty Images ; p. 249 Shutterstock.com; p. 251 Library of Congress; p. 253 Shutterstock.com; p. 256 Shutterstock.com

Copyright © 2004, 2018 by Black Dog & Leventhal, Inc.

Cover design by Carlos Esparza

Cover copyright © 2018 by Hachette Book Group, Inc.

Hachette Book Group supports the right to free expression and the value of copyright. The purpose of copyright is to encourage writers and artists to produce the creative works that enrich our culture.

The scanning, uploading, and distribution of this book without permission is a theft of the author's intellectual property. If you would like permission to use material from the book (other than for review purposes), please contact permissions@hbgusa.com. Thank you for your support of the author's rights.

Black Dog & Leventhal Publishers

Hachette Book Group

1290 Avenue of the Americas

New York, NY 10104

www.hachettebookgroup.com

www.blackdogandleventhal.com

Originally published in hardcover by Black Dog & Leventhal, Inc., May 2004

First revised edition: February 2018

Black Dog & Leventhal Publishers is an imprint of Hachette Books, a division of Hachette Book Group. The Black Dog & Leventhal Publishers name and logo are trademarks of Hachette Book Group, Inc.

The publisher is not responsible for websites (or their content) that are not owned by the publisher.

The Hachette Speakers Bureau provides a wide range of authors for speaking events. To find out more, go to www.HachetteSpeakersBureau.com or call (866) 376-6591.

Print book interior design by Clea Chmela

Library of Congress Control Number: 2017941395

ISBN: 978-0-316-43849-0 (paper over board); 978-0-316-43848-3 (ebook)

Printed in China

1010

10 9 8 7 6 5 4 3 2 1